ENERGY POLICIES, POLITICS AND PRICES

DEPARTMENT OF ENERGY FUNDED INVENTIONS

DISCLOSURE AND INTEREST PROTECTION ISSUES

ENERGY POLICIES, POLITICS AND PRICES

Additional books in this series can be found on Nova's website under the Series tab.

Additional e-books in this series can be found on Nova's website under the e-book tab.

ENERGY POLICIES, POLITICS AND PRICES

DEPARTMENT OF ENERGY FUNDED INVENTIONS

DISCLOSURE AND INTEREST PROTECTION ISSUES

PETER REED
EDITOR

NOVINKA

New York

Library of Congress Cataloging-in-Publication Data

ISBN: 978-1-63483-259-5

Published by Nova Science Publishers, Inc. † New York

CONTENTS

PREFACE

The U.S. Department of Energy (DOE) provides funding to contractors for research and development of new technologies. To incentivize participation in federal research projects and promote the use of federally funded inventions, the 1980 Bayh-Dole Act and other laws and regulations allow contractors receiving federal research and development funds to retain ownership of inventions they create so long as they adhere to certain requirements, including disclosing inventions developed with agency funding. DOE's ability to protect its interests in these inventions—including their utilization and domestic manufacture—depends on its knowledge of their existence. This book examines DOE funding for contractor research for fiscal years 2009 through 2013 and how DOE ensures that contractors disclose agency funded inventions; the challenges DOE faces in ensuring invention disclosure and actions it is taking to address them; and the challenges DOE faces in protecting its interests in these inventions and the actions it is taking to address them. Furthermore, the authors of this book review the policies and procedures used to determine whether march-in authority should be exercised in the Bayh-Dole Act; how the march-in authority has been used; and what barriers and disincentives have been encountered in exercising the march-in authority.

In: Department of Energy Funded Inventions ISBN: 978-1-63483-259-5
Editor: Peter Reed

Chapter 1

FEDERAL RESEARCH: DOE IS ADDRESSING INVENTION DISCLOSURE AND OTHER CHALLENGES BUT NEEDS A PLAN TO GUIDE DATA MANAGEMENT IMPROVEMENTS*

United States Government Accountability Office

ABBREVIATIONS

CFIUS	Committee on Foreign Investment in the United States
DOE	U.S. Department of Energy
iEdison	Interagency Edison
IP	Intellectual Property
PATMIS	Patent Management Information System

WHY GAO DID THIS STUDY

DOE provides funding to contractors for research and development of new technologies. To incentivize participation in federal research projects and promote the use of federally funded inventions, the 1980 Bayh-Dole Act and

* This is an edited, reformatted and augmented version of The United States Government Accountability Office publication, GAO-15-212, dated January 2015.

other laws and regulations allow contractors receiving federal research and development funds to retain ownership of inventions they create so long as they adhere to certain requirements, including disclosing inventions developed with agency funding. DOE's ability to protect its interests in these inventions—including their utilization and domestic manufacture—depends on its knowledge of their existence.

GAO was asked to review DOE efforts to protect its interests in agency funded inventions. This report examines: (1) DOE funding for contractor research for fiscal years 2009 through 2013 and how DOE ensures that contractors disclose agency funded inventions, (2) the challenges DOE faces in ensuring invention disclosure and actions it is taking to address them, and (3) the challenges DOE faces in protecting its interests in these inventions and the actions it is taking to address them. GAO reviewed laws, regulations, and other documents and interviewed DOE patent counsel responsible for intellectual property issues, representatives of organizations that facilitate the development of federally funded technology, and others.

What GAO Recommends

GAO recommends that DOE develop an implementation plan with milestones for improving its data management systems. DOE agreed with this recommendation.

What GAO Found

The U.S. Department of Energy (DOE) provided at least a total of $11 billion ($12 billion in fiscal year 2014 dollars) in research and development funding to contractors for fiscal years 2009 through 2013. Contractors reported about 5,800 inventions and 700 patents developed with DOE funding during this time period. To ensure disclosure of these agency funded inventions, DOE relies primarily on contractor self-reporting and financial assistance award closeout procedures. Contractors are generally required to adhere to specific time frames for invention disclosure. Following contractor invention disclosure, DOE patent counsel monitor the invention through the end of a financial assistance award to ensure contractor compliance with time frame requirements for electing to retain ownership and applying for patent protection of the invention.

DOE faces challenges in (1) ensuring that contractors disclose agency funded inventions and (2) managing information related to these disclosures and is taking steps to address them.

- *Limited ability to ensure invention disclosure after funding ends:* DOE does not have a documented process to ensure contractors disclose inventions after financial assistance awards end. To address this, DOE recently began two pilot efforts to determine the extent of undisclosed inventions. One is an audit of a sample of previously completed financial assistance awards and the other involves cross-referencing U.S. Patent and Trademark Office data against DOE information on inventions it funded. DOE is still implementing these efforts but reported identifying more than 100 potential undisclosed inventions. DOE will assess the results of the pilots to determine whether to continue them, according to DOE patent counsel.
- *Data management limitations:* DOE faces a challenge in managing information related to agency funded inventions because it relies on two different data systems that are outdated, unable to communicate with each other, and do not allow for electronic reporting. Under federal internal control standards, information should be recorded and communicated to management and others within the entity who need it and in a form and within a time frame that enables them to carry out their responsibilities. DOE is in the process of updating its data systems and is planning the development of an electronic reporting function but has not established an implementation plan with milestones against which it can track its progress toward completing these efforts. By developing such a plan, DOE would have greater assurance that it is making timely progress toward these efforts.

In addition, DOE faces challenges in its ability to monitor and influence the utilization and domestic manufacture of inventions it funded to protect its interests in them. DOE has proposed regulatory changes to address these challenges that would (1) require contractors to report on the utilization and domestic manufacture of agency funded inventions, (2) allow DOE to assess manufacturing plans as criteria for funding decisions, and (3) require contractors to obtain DOE authorization for changes in their control—

including ownership— under certain circumstances. According to patent counsel, DOE expects to finalize these regulatory changes in fiscal year 2015.

* * *

January 30, 2015

The Honorable Charles E. Grassley
Chairman
Committee on the Judiciary
United States Senate

The Honorable John R. Thune
United States Senate

Technological innovation is widely seen as responsible for much of the economic growth and increased standard of living in modern societies. Federal departments and agencies, such as the U.S. Department of Energy (DOE), support technological innovation through a wide range of research activities that focus on their mission needs. For example, DOE annually funds billions of dollars toward the nation's research in support of its mission to ensure America's security and prosperity by addressing its energy, environmental, and nuclear challenges through transformative science and technology solutions.

A large portion of DOE funded research is performed outside of the network of its 17 national laboratories through financial assistance awards with private sector or academic researchers, among others. The 1980 Bayh-Dole Act and other laws and implementing regulations provide the legal framework for the ownership of inventions developed by nonfederal recipients of federal research and development funding, commonly referred to as contractors.[1] To better transfer the results of federally funded research to the public, this framework allows contractors—rather than the government—to manage invention utilization by retaining ownership of inventions they develop with federal funds. At the same time, this framework requires contractors to take steps to protect the government's interests in these inventions. Specifically, DOE has an interest in facilitating the utilization of inventions resulting from the research it funds because the public benefits when technological advances become new goods and services in the marketplace. Additionally, DOE generally has an interest in federally funded inventions being made in the United States. To help protect these interests, the Bayh-Dole Act and DOE's large business patent waivers establish time frames in which contractors must disclose the development of federally funded inventions, determine whether to retain ownership, and apply for patent protection, among other things.[2] Contractor compliance with these requirements provides information on federally funded inventions that agencies can use to protect the government's interests in them.

In recent years, some congressional hearings raised concerns about whether inventions funded by agencies, such as DOE, are being used to support U.S. economic growth and competitiveness. For example, a December 2012 congressional hearing raised several concerns about whether taxpayer-funded research and development benefits foreign entities when such entities purchase U.S companies or their technology assets. [3] In addition, a March 2012 congressional hearing raised concerns about U.S. firms that relocate their manufacturing efforts overseas, particularly when federal agencies helped to fund the technologies used by those firms.[4]

You asked us to review DOE's efforts to protect its interests in agency funded inventions. This report examines (1) the extent to which DOE funds contractor research and how it ensures that contractors disclose inventions they develop with agency funding; (2) the challenges DOE faces in ensuring contractors disclose inventions, as well as the actions it is taking to address them; and (3) the challenges DOE faces in protecting its interests in agency funded inventions, as well as the actions it is taking to address them.

To examine how DOE ensures that contractors disclose inventions they develop with agency funding, we identified and reviewed laws, regulations, and agency documents associated with invention disclosure requirements and DOE's process to monitor contractor compliance with these requirements.[5] We also interviewed DOE patent counsel responsible for monitoring contractor compliance with invention disclosure requirements.[6] To examine the extent to which DOE funds contractor research, we collected and analyzed data on the extent to which DOE funded research by contractors subject to the provisions of the Bayh-Dole Act and other relevant laws and regulations from fiscal years 2009 through 2013.[7] Additionally, we collected information from DOE on the number of contractor-disclosed inventions and patents during this same time period. We reviewed DOE's practices for recording this data and information and determined they were sufficiently reliable for our purposes.[8] To examine the challenges that DOE faces in ensuring that contractors disclose inventions they develop with agency funding and any actions it is taking to address them, we reviewed applicable regulations, as well as DOE processes, procedures, and data systems for monitoring contractor compliance. We also reviewed documentation of DOE efforts to address the challenges it faces, and we interviewed DOE patent counsel. To examine the challenges that DOE faces in protecting its interests in agency funded inventions as well as the actions it is taking to address them, we reviewed laws, regulations, and agency documents associated with DOE's interests in these inventions and interviewed DOE patent counsel. To understand issues associated with the development and patenting of federally funded inventions that underlie our objectives, we interviewed officials from the U.S. Patent and Trademark Office, representatives from several industry organizations involved in facilitating the development of federally funded technology, and academic researchers who study the topic.

BACKGROUND

We conducted this performance audit from October 2013 to January 2015 in accordance with generally accepted government auditing standards. Those standards require that we plan and perform the audit to obtain sufficient, appropriate evidence to provide a reasonable basis for our findings and conclusions based on our audit objectives. We believe that the evidence obtained provides a reasonable basis for our findings and conclusions based on our audit objectives.

Prior to the Bayh-Dole Act, the government generally retained ownership of federally funded inventions regardless of whether the research was performed in federal laboratories, at universities, or by individual companies, and only 5 percent of the patents on these inventions were ever used in the private sector, according to a Congressional Research Service report.[9] The Bayh-Dole Act allows nonprofits, small businesses, and universities to retain ownership of federally funded inventions to promote the utilization of inventions created through federal research and development programs, and to provide an incentive for contractors to commercialize federally funded inventions for sale in the marketplace. The Bayh-Dole Act applies to universities, nonprofit organizations, and small businesses that receive federal research funding. Federal agencies that enter into financial assistance awards with these types of entities must adhere to the requirements of the Bayh-Dole Act and its implementing regulation developed by the Department of Commerce.[10] In particular, the Bayh-Dole Act requires agencies, including DOE, to incorporate specific provisions into research and development financial assistance awards that contractors must comply with to help protect the government's interests in any resulting inventions.

Additional laws and implementing regulations allow DOE to grant waivers to large businesses—if petitioned to do so—so that they will be able to retain ownership of inventions they develop under financial assistance awards. Specifically, the Federal Nonnuclear Energy Research and Development Act of 1974 [11] and the Atomic Energy Act of 1954[12] provide DOE with the authority to waive its right to ownership of inventions developed with federal funding by contractors. Additionally, a presidential memorandum in 1983, followed by an executive order in 1987, directed federal agencies, to the extent permitted by laws such as these, to establish policies that are substantially the same as those contained in the Bayh-Dole Act for all businesses. To do so, DOE uses its Patent Waiver Regulation to grant waivers to large businesses so that they may retain ownership of inventions they develop with agency funding.[13]

DOE generally includes specific provisions in its financial assistance awards that require contractors to disclose inventions stemming from agency funding to the government and to meet certain time frames for taking action to protect these inventions. For example, according to these provisions, small businesses, nonprofits, and universities must

1. disclose any invention that results from agency funding within 2 months after learning the invention has been created,
2. notify the funding agency as to whether they elect to retain ownership of an invention within 2 years after they disclose it, and
3. apply for a patent on an invention, typically within 1 year of electing to retain ownership.

DOE generally requires similar actions from large businesses in order to retain ownership of agency funded inventions, but the time frames can vary depending on the specific provisions of an individual financial assistance agreement.

Compliance with these provisions informs agencies, such as DOE, of an invention's existence, and helps ensure that a contractor takes timely steps to patent the intellectual property embodied in the invention, should the contractor wish to retain ownership of it. If a contractor does not comply with these requirements, agencies have the authority to demand ownership of a federally funded invention to provide the government the opportunity to take steps to protect its interests in it. Once contractors own a patented invention, they are generally free to use or transfer their patent rights at their discretion, in compliance with applicable laws, such as those regarding export control.[14]

The Bayh-Dole Act also identifies certain interests the government has in federally funded inventions, including their utilization and domestic manufacture. DOE's Patent Waiver Regulation does not explicitly identify these interests with regard to inventions developed by large business contractors, but the waivers themselves generally will outline specific requirements related to invention utilization and domestic manufacture as shown below.

- *Utilization.* DOE can use the authority, known as march-in authority, to require a contractor or licensee to grant a license to any responsible entity or entities when an agency determines that certain conditions identified in the act have been met.[15] Additionally, the government has a nonexclusive royalty-free license to practice, and have practiced on its behalf, any invention created with federal funding.
- *Domestic manufacture.* Under the Bayh-Dole Act and DOE regulations and policy, DOE has established certain requirements in financial assistance awards for contractors to manufacture federally funded inventions in the United States. These requirements vary depending on the type of contractor involved in a financial assistance award.[16] Domestic manufacture requirements for federally funded inventions include:
  - *U.S. Preference.* U.S. Preference provisions generally require small business and nonprofit contractors' exclusive licensees to substantially manufacture federally funded inventions domestically in order to use or sell the inventions in the United States.[17] These requirements apply only to a contractor's exclusive licensee; the contractor that developed the invention faces no limitations on manufacturing location. Federal agencies may use their march-in authority in instances where contractors' licensees do not comply with U.S. Preference requirements.
  - *U.S. Competitiveness.* U.S. Competitiveness provisions generally require that the manufacture of inventions developed by large businesses must occur substantially in the United States unless otherwise approved by DOE. DOE may require forfeiture of invention ownership or refund of its investment when contractors do not comply with these requirements if such penalties are

included in the patent rights clause of the financial assistance award.

- *U.S. Manufacturing Plan.* U.S. Manufacturing Plan provisions generally require that contractors submit plans as part of a funding application specifying how they intend to domestically manufacture any potential inventions developed in the course of the financial assistance award.[18] Where this provision is part of a financial assistance award, DOE may take ownership of the invention, require refund of its investment,[19] or establish additional reporting requirements in instances of noncompliance.

In addition, DOE may include provisions within a financial assistance award that require reports from contractors to provide information on how federally funded inventions are utilized.[20] Specifically, DOE programs can request that contractors provide periodic reports with information regarding patent status (e.g., filing date, application number and title, and patent number and issue date), and invention utilization (e.g., the status of development, date of first commercial sale or use, and gross royalties received). DOE may request these reports at its discretion but not more than annually.

DOE Relies on Contractor Self-Reporting and Agency Procedures to Help Ensure Disclosure of Agency Funded Inventions

During fiscal years 2009 through 2013, DOE initiated nearly 6,000 financial assistance awards. DOE relies primarily on contractor self-reporting and financial assistance award closeout procedures to ensure that contractors disclose agency funded inventions.

During Fiscal Years 2009 through 2013, DOE Initiated Nearly 6,000 Financial Assistance Awards

During fiscal years 2009 through 2013, DOE initiated a total of nearly 6,000 financial assistance awards worth at least $11 billion with contractors, according to data provided by DOE.[21] During this period, according to DOE patent counsel, contractors reported approximately 5,800 inventions, elected to take ownership of about 2,800 inventions, and were issued more than 700 patents.[22] Figures 1 and 2 show the amount and number of financial assistance awards DOE established with different types of contractors for fiscal years 2009 through 2013. In fiscal year 2013, the most recent year data were available, DOE provided a majority of funds to universities and nonprofits (nearly $1 billion across 580 agreements), followed by large businesses (approximately $304 million across 79 agreements) and small businesses (approximately $290 million across 370 agreements).

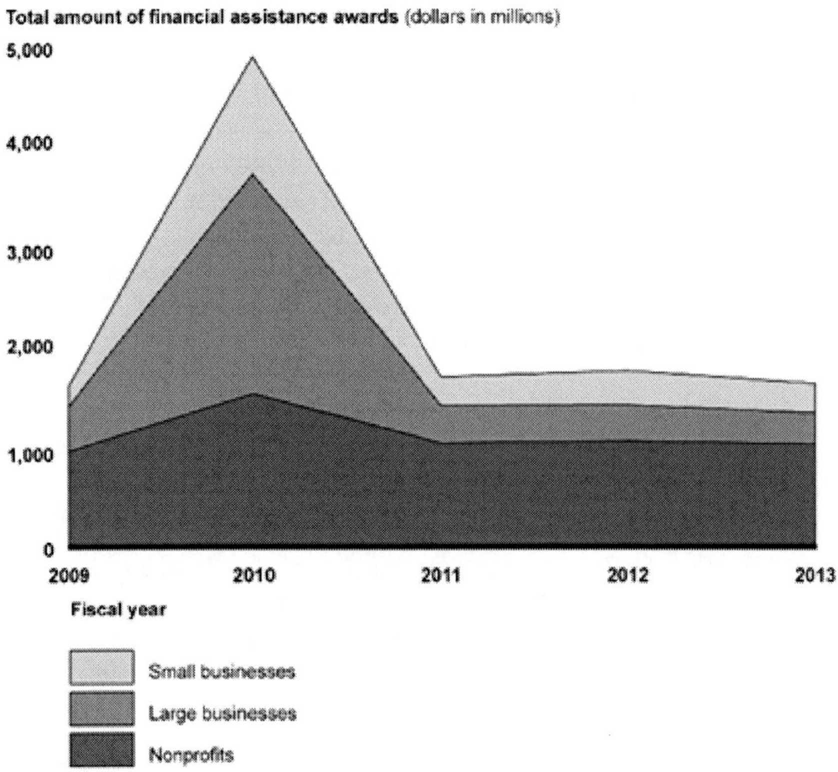

Source: GAO analysis of Department of Energy data. | GAO-15-212.

Figure 1. Amount of Department of Energy Financial Assistance Awards, by Contractor Type, Fiscal Years 2009-2013.

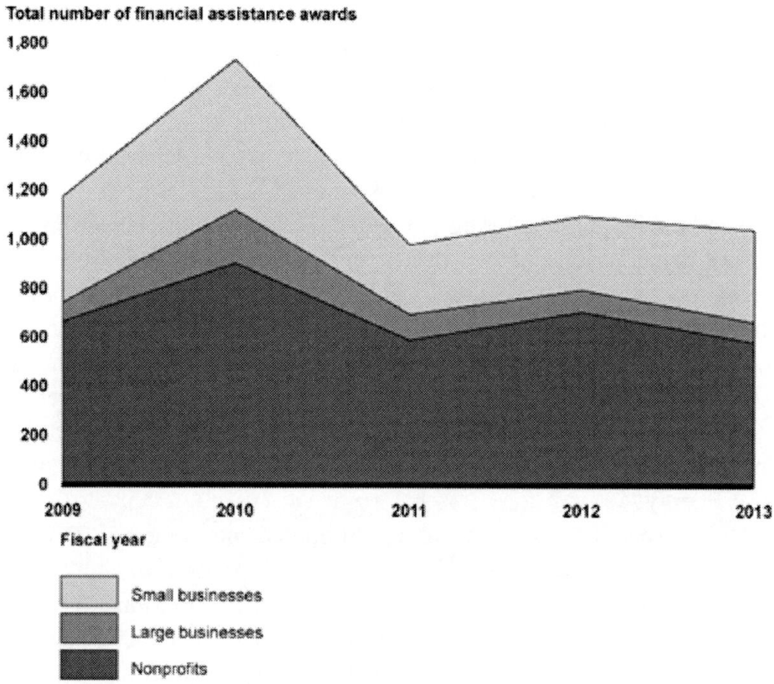

Source: GAO analysis of Department of Energy data. | GAO-15-212.

Figure 2. Number of Department of Energy Financial Assistance Awards, by Contractor Type, Fiscal Years 2009-2013.

DOE Relies on Contractor Self-Disclosure and Agency Procedures to Help Ensure Disclosure of Agency Funded Inventions

Contractors disclose agency funded inventions to DOE through a variety of reporting mechanisms including e-mail, regular mail, and Interagency Edison (iEdison)—which is an electronic reporting system that allows federal grantees and contractors to report federally funded inventions, patents, and utilization data.[23] According to DOE procedures, when a contractor discloses an invention developed with DOE funds, patent staff create a file for that invention. According to DOE procedures, patent counsel then monitor compliance with time frames related to the contractor's determination of whether to retain ownership and pursue patent protection. From this point forward, according to DOE procedures, whenever a contractor submits additional information about an invention, DOE patent counsel review the invention's file to ensure compliance with all invention disclosure provisions. If DOE patent counsel determine that a contractor has not met the specified time frames for disclosing and electing ownership of an invention, they send the contractor a letter— known as a demand letter—demanding that the contractor give DOE ownership of the invention.[24]

To assist with the monitoring efforts, DOE patent counsel enter invention disclosure data into one of two data systems: (1) Intellectual Property (IP) Manager,[25] which patent counsel said they use to manage information regarding inventions funded by the Advanced Research Projects Agency-Energy and the National Nuclear Security Administration, and (2) Patent Management Information System (PATMIS), which patent counsel said they use to manage information about inventions funded by DOE's Offices of Science, Energy Efficiency and Renewable Energy, Fossil Energy, and Nuclear Energy.[26]

DOE has specific closeout procedures it implements at the end of financial assistance awards, and DOE officials said these procedures are the main compliance check to ensure that agency funded inventions have been disclosed. DOE's Financial Assistance Rules state that a contractor has up to 3 months from the conclusion of a financial assistance award to submit all documents required to meet the terms of the award. Then, during closeout, DOE patent counsel conduct a final compliance check for invention disclosures and other required documentation. DOE patent counsel stated that they discover the vast majority of contractor noncompliance with invention disclosure requirements during closeout.[27] They added that nondisclosure, particularly from contractors with more limited resources, is generally inadvertent. They told us that nondisclosure is more often due to limited contractor familiarity with the legal requirements or experience in working with the federal government than an intentional attempt to avoid disclosing an agency funded invention. DOE patent counsel said if a contractor failed to disclose an agency funded invention but did so inadvertently, they would generally work with the contractor to meet disclosure requirements rather than demand ownership for noncompliance with disclosure requirements. They said that approach better meets their view of the intent of the Bayh-Dole Act to commercialize federally funded inventions, while ensuring the government protects its interests in them.

The closeout of a financial assistance award is also the point when DOE monitors whether contractors have patented agency funded inventions. For example, according to DOE procedures, a contractor must send DOE patent counsel a patent certification form—a document that discloses all inventions that resulted from the financial assistance award. In turn, according to DOE procedures, patent counsel use this information to ensure contractor compliance with the terms of the financial assistance award before issuing a patent clearance letter—a document in which DOE acknowledges the contractor's ownership of the patented invention. In addition to the required certification form, upon request from DOE, contractors must provide information about patents for any invention they elected to own. When a contractor submits such information, DOE patent counsel verify that the reported patents include a statement of government interest—standard language in a patent that acknowledges the government's role in funding its development—and that the invention's file is complete and up-to-date in the relevant data system.

DOE Faces Challenges in Ensuring that Contractors Disclose Agency Funded Inventions and Is Taking Actions to Address Them

DOE faces challenges in ensuring that contractors disclose inventions they develop with agency funding and is taking actions to address them. Specifically, one challenge DOE faces is not having a documented process for ensuring that contractors disclose agency funded inventions after financial assistance awards end, and the agency has initiated two pilot projects to identify the extent of potentially undisclosed inventions. Additionally, DOE faces a challenge in managing invention disclosure information because its data systems for doing so have limited capabilities. While the agency has begun to upgrade those systems, DOE has not developed an implementation plan with specific milestones for certain key steps to guide these efforts.

DOE Is Implementing Pilot Projects to Address Challenges in Identifying Undisclosed Inventions

One challenge that DOE faces is not having a documented process for ensuring that contractors disclose agency funded inventions after financial assistance awards end. DOE's closeout procedures are designed to, among other things, ensure that contractors disclose all inventions developed during a

DOE financial assistance award and that DOE's interests in them are documented. However, DOE has no documented process to monitor inventions after award closeout. Due to the time that may elapse between when a contractor files a patent application and when a patent is granted by the U.S. Patent and Trademark Office, DOE funded inventions may exist that are not identifiable through public searches at the time a financial assistance award ends. Under such a scenario, if the contractor did not voluntarily disclose an invention, the invention would not be identifiable during DOE's closeout procedures. Instead, DOE patent counsel told us they rely on contractors to voluntarily disclose such information following financial assistance award closeout. As a result, the potential exists for DOE to be unaware of inventions that it funded and in which it would retain interests.

To address this challenge, DOE recently launched two pilot projects aimed at better understanding the extent of undisclosed inventions. One pilot project is an audit of a sample of previously completed financial assistance awards to determine the extent to which contractors did not disclose DOE funded inventions. The second pilot project involves cross-referencing U.S. Patent and Trademark Office data against DOE information on inventions it funded.

In the first pilot project, which began in December 2013, DOE developed draft audit procedures to sample previously completed financial assistance awards and determine the extent to which contractors did not disclose agency funded inventions. Under the draft audit procedures, DOE patent counsel (1) audit contractor activities for any indication of patents, patent applications, or inventions that might have been developed with DOE funding but were not disclosed to DOE; (2) submit demand letters to contractors to obtain ownership of any potentially undisclosed inventions; and (3) work with contractors, depending on the circumstances of nondisclosure, to allow them to retain ownership of the inventions.[28] The draft audit procedures set a target of annually sampling 100 randomly selected financial assistance awards completed during the previous 5 years, which represents approximately 5 percent of the agreements closed out during that period, according to DOE documentation. As of December 2014, DOE patent counsel told us that they had reviewed 99 financial assistance awards completed during the previous 5 years and identified three undisclosed inventions for which they sent demand letters to the relevant contractors. DOE patent counsel told us that they granted one of the contractors an extension of time to retain ownership of the undisclosed invention because DOE determined that the nondisclosure was an oversight. DOE patent counsel said that they resolved another as a data entry error. As of December 2014, DOE had not received a response to the third demand letter. DOE modified its draft audit procedures based on initial testing, and patent counsel told us that they might conduct another phase of pilot testing before assessing the overall results of the audit procedures and making a determination about whether to implement the pilot project on a permanent basis. They told us that this decision will depend on the extent to which the pilot project identifies undisclosed inventions compared with the resources—principally staff hours—necessary to conduct it. For example, if the audit procedures identify few undisclosed inventions, DOE could be less likely to implement it permanently.

DOE's second pilot project involves cross-referencing U.S. Patent and Trademark Office data against DOE information on inventions it funded. For this pilot project, DOE patent counsel analyzed data from a U.S. Patent and Trademark Office database to identify inventions or patents with a government interest clause indicating the patent stemmed from DOE funding. Then, DOE patent counsel reviewed the data to determine if information on issued patents or published patent applications also existed in DOE's PATMIS data system, with patent files not identified in DOE's data system representing possible unreported inventions or patents. For the pilot, DOE patent counsel told us that they reviewed 549 patented inventions that they identified as absent from DOE's data system. Through this effort, DOE patent counsel told us they identified 100 patented inventions that may have been undisclosed and issued 40 demand letters covering 64 undisclosed inventions to contractors.[29] As of December 2014, they said DOE resolved 16 demand letters—covering 22 inventions—and is awaiting response from the other 24—covering the remaining 42 inventions. According to DOE patent counsel, 11 of the contractors had not properly disclosed the patented inventions, generally due to misunderstandings regarding what their responsibilities were to the government. They told us that, in each instance, DOE determined that the contractor had acted in good faith and allowed an extension of time for the contractors to file the required disclosure paperwork to retain ownership. According to DOE patent counsel, the 5 other contractors to which DOE issued demand letters demonstrated that they had not failed to comply with invention disclosure requirements. DOE patent counsel explained that the related inventions were either disclosed to DOE but not logged into its data systems, or were logged incorrectly into its data systems, among other reasons. DOE patent counsel said that this effort helped DOE identify several inventions that were not properly disclosed. DOE patent counsel also indicated that the pilot revealed that the contractors' failure to report inventions to DOE appeared to have been in good faith, but that the contractors were unaware of all invention reporting obligations. DOE patent counsel told us that, in turn, the agency is reviewing the compliance assistance resources available to contractors and periodically reminds them of their obligations to DOE. They also said that, after these compliance assistance actions have been implemented, DOE anticipates reevaluating the usefulness of the pilot audit project.

DOE Is Planning to Update Data Systems to Address Challenges in Managing Invention Disclosure Information but Does Not Have an Implementation Plan with Key Milestones

DOE faces a challenge in managing information on disclosed inventions, which underpins its ability to protect its interests in agency funded inventions. This is because, according to DOE patent counsel, the department's two older invention data management systems—IP Master and PATMIS—are outdated, unable to communicate with each other, and do not have functionality for electronically updating invention disclosure or patent status, hampering their ability to manage information and data related to inventions developed with DOE funding. For example, DOE patent counsel told us that IP Master will soon be obsolete because, among other reasons, the vendor is discontinuing support for it. Additionally, DOE patent counsel told us that the existence of two separate systems can lead to duplicative entries and make it difficult to track invention disclosures on a department-wide basis. Further, DOE patent counsel explained that the systems do not have the functionality for electronic reporting. That means contractors submit information—such as invention disclosure reports—in mail, faxes, and e-mails, and DOE patent counsel must then manually enter that information in DOE's systems. The DOE patent counsel described this as a time-consuming, labor-intensive process that can increase the likelihood of data errors.

DOE patent counsel stated that they recognized the need to transition to a unified data tracking and monitoring system with functionality for electronic reporting and have begun efforts to do so. Specifically, DOE patent counsel told us that they recently upgraded one of the agency's data management systems—IP Master—to a new system called IP Manager, which became operational in November 2014. DOE patent counsel said they intend to migrate the agency's other system— PATMIS—to the new IP Manager system in the near future so there will be only one data tracking and monitoring system. However, prior to any such migration of PATMIS, DOE officials told us that they want to ensure that there are no technical problems with the new IP Manager system and evaluate its performance in case modifications are necessary.

DOE patent counsel told us that integrating DOE's existing systems into the IP Manager system will provide a consolidated, DOE-wide invention management database that will significantly reduce administrative costs associated with invention management and improve data quality by reducing duplicative entry of inventions. Also, according to documentation DOE provided about the IP Manager system, some of its features may help address DOE's current data management challenges including automated generation of legal forms to reduce the need to draft unique documentation for every invention. Additionally, DOE plans to develop a "one click invention reporting" capability that would be designed to, among other things, give contractors the ability to electronically report and update invention records directly from their own databases and provide DOE with enhanced reporting functions to monitor contractor actions, including invention disclosure.[30] DOE patent counsel explained that this would enhance DOE's ability to share invention information across the agency to better enable it to manage information to track contractor compliance with financial assistance award provisions. DOE patent counsel told us the agency has created a workgroup composed of representatives from federal agencies and DOE laboratories to determine the requirements for this new "one click invention reporting" capability and tentatively plans to begin software development in the spring of 2015, with an initial pilot capability scheduled for fall of 2015 and full deployment in early 2016, subject to available funding.

While DOE transitioned its IP Master system to IP Manager, and has developed an implementation plan with milestones for the additional capability it wants to add to the IP Manager system, DOE patent counsel said the agency does not have an implementation plan with milestones for when PATMIS would be migrated to IP Manager. They said that PATMIS contains the majority of the agency's invention records and estimated that, given the amount of data contained in the system, it could cost as much as $100,000 to move the data from PATMIS into IP Manager. According to DOE patent counsel, as of November 2014, DOE had not prioritized funding to migrate the data and consolidate the databases. DOE also has not established milestones for its efforts to (1) identify requirements for the "one click invention reporting" capability it plans to add to IP Manager or (2) evaluate any system modifications necessary as a result of migrating IP Master to IP Manager and associated system requirements.

Under federal standards for internal control, information should be recorded and communicated to management and others within the entity who need it and in a form and within a time frame that enables them to carry out their internal control and other responsibilities.31 In addition, best practices for information technology system acquisition emphasize developing requirements, among other practices, to guide software engineering investments.[32] By planning to transition to a unified data tracking and monitoring system with functionality for electronic reporting, DOE is taking a step in the right direction. Also, the implementation plan with milestones DOE developed for the additional capability it wants to add to the IP Manager system will help it track and communicate progress toward completing this effort. By developing a comprehensive implementation plan with milestones that cover all aspects of the steps DOE is taking to transition to a unified data tracking and monitoring system with functionality for electronic reporting, DOE would have greater assurance that it will be able to track progress toward completing all of these steps in a timely manner.

DOE Proposed Regulatory Changes to Address Challenges to Monitoring and Influencing Contractor Utilization and Domestic Manufacture of Agency Funded Inventions

DOE faces challenges in monitoring and influencing contractor utilization and domestic manufacture of agency funded inventions to protect its interests in them, and the agency has proposed regulatory changes to address these challenges.[33] Specifically, DOE does not have a standard for invention utilization and manufacture reporting, has a limited ability to compel domestic manufacture of agency funded inventions, and has a limited ability to influence changes in control of contractors receiving DOE funds. In turn, DOE's proposed regulatory changes address, with respect to for-profit contractors, the frequency and duration of invention utilization and manufacturing reporting, use of U.S. Manufacturing Plans, and DOE influence over changes in contractor control.[34] DOE patent counsel indicated that the agency anticipates issuing a final rule to implement these changes in fiscal year 2015.

DOE Does Not Have a Standard for Invention Utilization and Manufacture Reporting

DOE faces a challenge in understanding how inventions it funds are utilized and manufactured because it does not have a standard for collecting such information from contractors. Currently, DOE can, at its discretion, request such information, but DOE regulations do not specify the duration of this reporting. DOE patent counsel told us that there are no established procedures regarding the frequency and duration of requests for information on invention utilization and manufacturing of agency funded inventions, which could hamper DOE's efforts to protect its interests. To address this challenge, DOE's proposed regulatory change would require annual contractor reporting on the utilization and manufacture of any products that use a DOE funded invention.[35] This reporting would include information on the status of technology development, date of first commercial sale or use, gross royalties received, and manufacturing locations of those products, and it would be required for at least 10 years following invention disclosure. DOE patent counsel said that requiring this reporting for all DOE financial assistance awards to for-profit contractors would enhance the agency's ability to protect its interests by providing it additional information with which to monitor compliance with financial assistance award provisions.

DOE faces a challenge in protecting the agency's interest in the domestic manufacture of agency funded inventions. DOE patent counsel stated that the agency's ability to compel domestic manufacture of agency funded inventions is limited because the agency can only require substantial domestic manufacture from certain contractors under certain circumstances. In particular, they said that the process for determining an "exceptional circumstance"—which DOE can use to broaden the domestic manufacture requirements beyond those included in award provisions— can be difficult. Specifically, they noted that it is not a routine process and often requires substantial internal agency analysis, as well as coordination with other agencies, including the Department of Commerce. Additionally, any broadened requirements apply only to the specific programs or activities covered by the determination.[36] For example, in 2011, DOE determined that an exceptional circumstance existed regarding DOE's SunShot Initiative.[37] Specifically, DOE proposed requiring a certain level of domestic manufacture for all inventions developed under the SunShot Initiative by all contractors, as well as their licensees.[38] However, this proposed requirement was not implemented following a review to identify international trade and other issues, according to DOE patent counsel. DOE patent counsel also said that the limitations of U.S. Preference provisions—which apply to small businesses, academia, and nonprofits—constitute a challenge to protecting the agency's interest in domestic manufacture. Specifically, such provisions require the exclusive licensees of contractors, but not the contractors themselves, to substantially manufacture agency funded inventions domestically. To address this limitation, DOE's proposed regulatory change would allow programs to require that potential for-profit contractors submit a U.S. Manufacturing Plan as part of any funding proposal.[39] In turn, DOE would be able to use these plans as criteria in assessing funding proposals and making financial assistance award decisions. Under the proposed regulatory change, and consistent with applicable law, DOE could make such plans binding for any licensee or entity that subsequently acquired ownership of an invention or technology developed under a DOE financial assistance award. DOE patent counsel explained that requiring and enforcing U.S. Manufacturing Plans increases the agency's ability to influence the domestic manufacture of inventions it funded.

DOE Has Limited Ability to Influence Changes in Control of Contractors

DOE faces a challenge in influencing the change in control—such as ownership—of contractors engaged in DOE financial assistance awards. A change in control may affect a contractor's ability to carry out the project that DOE is funding and, according to DOE patent counsel, the agency currently does not require contractors to notify it of such changes. DOE's proposed regulation would require contractors to notify the agency of changes in their control in certain circumstances.[40] The proposed regulation would establish procedures for the change of control, including ownership, of a contractor that received a DOE funding award of more than $10 million. The affected contractor would be required to notify the agency within 30 days of a change of control, or within 30 days if the contractor had a reason to know that such a change is likely.[41] Failure to notify DOE would be grounds for suspension or termination of the financial assistance award. Further, without DOE authorization of the change of control, the award funding could cease to continue.[42] However, such penalties would only apply to the current award.

CONCLUSION

DOE is taking actions to update its data systems and move to a consolidated, DOE-wide invention management database to improve its ability to monitor and manage information regarding contractor disclosure of agency funded inventions. Currently, its two data systems are outdated, unable to communicate with each other, and do not have functionality for electronically updating invention disclosure or patent status, hampering DOE's ability to manage information and data on agency funded inventions. DOE recently upgraded one of its data systems—IP Master—to a new system called IP Manager, which became operational in November 2014 and has developed an implementation plan with milestones for the additional capability it wants to add to the IP Manager system. However, DOE has not developed an implementation plan with appropriate milestones to help assess its progress toward completing key steps, such as moving the data in its primary PATMIS database—which contains the majority of the agency's invention records—into the new IP Manager system, or for defining the requirements for planned or potential upgrades to that system. Without such a plan, DOE may not have assurance that it is making timely progress toward obtaining the information management capabilities it needs to protect its interests in agency funded inventions by monitoring information regarding contractor disclosure of these inventions.

RECOMMENDATION FOR EXECUTIVE ACTION

To help provide greater assurance that DOE is making timely progress toward obtaining the information management capabilities it needs to protect its interests by monitoring contractor disclosure of agency funded inventions, we recommend that the Secretary of Energy develop an implementation plan, including appropriate milestones, to guide DOE's efforts to improve its data management capabilities.

AGENCY COMMENTS

We provided a draft of this report to the Department of Energy and Department of Commerce for review and comment. In its written comments, the Department of Energy agreed with our findings and recommendation. The Department indicated that it would develop an implementation plan to guide data management improvements. The Department of Commerce neither agreed nor disagreed with our findings but noted that it provided background information on intellectual property issues related to the review. The Department of Energy's and Department of Commerce's written comments are reproduced in appendixes I and II, respectively. Both agencies also provided technical comments that we incorporated, as appropriate.

John Neumann
Director,
Natural Resources and Environment

End Notes

[1] Under the Bayh-Dole Act, the term "contractor" means any person, small business firm, or nonprofit organization that is a party to a federal funding agreement, which includes contracts, grants, or cooperative agreements for the performance of experimental, developmental, or research work. 35 U.S.C. § 201(b)-(c). Section 9 of the Federal Nonnuclear Energy Research and Development Act of 1974 (42 U.S.C. § 5908), Section 152 of the Atomic Energy Act of 1954 (42 U.S.C. § 2182), and DOE's Patent Waiver Regulations (10 C.F.R. pt. 784), allow DOE to provide waivers to large businesses so that they may retain ownership of inventions developed with agency funding. For the purposes of this report, the term "contractor" includes large businesses receiving such waivers, as well as entities defined as contractors under the Bayh-Dole Act except when referring specifically to a contractor under a particular statute. In addition, this report does not address DOE management of all types of funding agreements, but only a subset of those—financial assistance awards.

[2] A U.S. patent is an exclusive right granted, generally for 20 years, within the United States to someone who invents or discovers (1) a new and useful process, machine, manufacture, or composition of matter or (2) any new and useful improvement of such items. A patent owner can grant permission to use a patented invention by licensing others to use, make, or sell the patented invention on an exclusive or nonexclusive basis. The U.S. Patent and Trademark Office is the federal agency responsible for granting U.S. patents and registering trademarks to carry out Congress' authority under Article I, Section 8, Clause 8, and Article I, Section 8, Clause 3, of the Constitution.

[3] The Impact of International Technology Transfer on American Research and Development, Before the House Committee on Science, Space, and Technology, Subcommittee on Investigations and Oversight, 112th Congress, 3-10 (2012).

[4] Commerce, Justice, Science, and Related Agencies Appropriations for 2013, Before the House Committee on Appropriations, Subcommittee on Commerce, Justice, Science, and Related Agencies, 112th Congress, 364-366 (2012).

[5] In reviewing how DOE ensures contractors disclose inventions developed with agency funding, we included activities related to DOE's civilian program offices but excluded the National Nuclear Security Administration given potential security and classification issues, as well as certain Bayh-Dole provisions that are specific to the activities of that agency.

[6] DOE patent counsel are located at various sites throughout the country and provide intellectual property law support to DOE's programs.

[7] We last reviewed the government's management of its interests in federally funded inventions in 2009, so this review focuses on the period since the issuance of our last report. See GAO, Federal Research: Information on the Government's Right to Assert Ownership Control over Federally Funded Inventions, GAO-09-742 (Washington, D.C.: July 27, 2009).

[8] DOE indicated that the potential exists for contractor underreporting and duplicative DOE entries in the systems it uses to monitor invention disclosure, but we determined these data are suitable for presenting approximate data on the number of inventions developed and patents issued for fiscal years 2009 through 2013.

[9] Congressional Research Service, The Bayh-Dole Act: Selected Issues in Patent Policy and the Commercialization of Technology, RL32076 (Washington, D.C.: Apr. 3, 2008).

[10] The Bayh-Dole Act directs the Department of Commerce to develop regulations to implement the provisions contained in the act. DOE also has regulations—DOE Financial Assistance Rules (see 10 C.F.R. pt. 600)—that provide the administrative requirements and operational rules for financial assistance award and administration.

[11] Federal Nonnuclear Energy Research and Development Act of 1974 § 9, 42 U.S.C. § 5908.

[12] Atomic Energy Act of 1954 § 152, 42 U.S.C. § 2182. 1310 C.F.R. pt. 784.

[14] The government retains certain rights and interests in these inventions even if they are subsequently assigned, licensed, or otherwise transferred to others.

[15] A licensee, in this context, is an entity that gains a legal license to exercise a patent for an invention from its owner (i.e., a contractor). DOE has not exercised its march-in authority because, officials explained, the agency has not discovered or received information indicating that circumstances exist in which use of this authority might be warranted.

[16] DOE may waive domestic manufacturing requirements if contractors or licensees show that reasonable but unsuccessful efforts have been made to substantially manufacture an invention in the United States or that, under existing circumstances, domestic manufacture is not commercially feasible.

[17] The Bayh-Dole Act and its implementing regulations do not define substantial domestic manufacturing.

[18] Under the Bayh-Dole Act, DOE may apply these provisions in instances where it determines that "exceptional circumstances" exist. 35 U.S.C. § 202(a)(ii). The Bayh-Dole Act provides the authority to make this determination in instances where certain additional or modified requirements in financial assistance awards would better promote the policy and objectives of the act. See id.

[19] DOE officials told us that the agency has not sought recoupment of agency funds for any breach of domestic manufacturing provisions.

[20] 37 C.F.R. § 401.8. DOE's large business patent waivers may contain similar reporting requirements.

[21] At least $12 billion in fiscal year 2014 dollars.

[22] DOE patent counsel provided information on the number of contractor reported, owned, and patented inventions based on the information contained in DOE information systems for tracking agency funded inventions.

[23] According to DOE patent counsel, contractors also, though infrequently, make such disclosures to DOE officials associated with managing the program through which they are receiving funds, and these program officials then forward such disclosures to DOE patent counsel.

[24] After a contractor receives a demand letter, it may request an extension to report the invention or provide evidence to DOE patent counsel that the invention in question is not subject to provisions of the financial assistance award, such as if the invention were developed without federal funding.

[25] DOE patent counsel told us that they recently upgraded one of the agency's data management systems—IP Master—to a new system called IP Manager, which became operational in November 2014.

[26] In 2007, the America Creating Opportunities to Meaningfully Promote Excellence in Technology, Education, and Science Act (Pub. L. No. 110-69, § 5012 (2007)) established the Advanced Research Projects Agency-Energy within DOE to overcome the long-term and high-risk technological barriers in the development of energy technologies. See GAO, Department of Energy: Advanced Research Projects Agency-Energy Could Benefit from Information on Applicants' Prior Funding, GAO-12-112 (Washington, D.C.: Jan. 13, 2012).

[27] DOE patent counsel said that the agency does not track the overall number of undisclosed inventions it identifies during the course of financial assistance awards.

[28] DOE's draft audit procedures identify circumstances—including written certification from contractors that any omission to disclose inventions was either in good faith or unintentional—under which DOE may grant extensions of time to contractors so that they may submit the necessary documentation to retain ownership.

[29] According to DOE patent counsel, one of the potentially undisclosed inventions resulted from work at Los Alamos National Laboratory and was handled by the National Nuclear Security Administration. As of December 2014, DOE was in the process of determining how to proceed with regard to the other 35 potentially undisclosed inventions.

[30] DOE intends for this software capability to be usable by any governmental agency that funds contractor-developed inventions. Therefore, DOE is including stakeholders from other governmental agencies in its development efforts.

[31] GAO, Standards for Internal Control in the Federal Government, GAO/AIMD-00-21.3.1 (Washington, D.C.: November 1999).

[32] See, for example, Carnegie Mellon Software Engineering Institute, Capability Maturity Model® Integration for Development (CMMI-DEV), Version 1.3 (November 2010). The Software Engineering Institute is a federally funded research and development center operated by Carnegie Mellon University. Its mission is to advance software engineering and related disciplines to ensure the development and operation of systems with predictable and improved cost, schedule, and quality.

[33] 79 Fed. Reg. 27,795 (May 15, 2014). The proposed regulations apply only to for-profit contractors. The public comment period for DOE's proposed regulatory changes ended in July 2014.

[34] Change in control of contractors includes the ownership of the contractor.

[35] The proposed regulations would allow DOE, at its discretion, to require more frequent than annual contractor reporting.

[36] In September 2013, DOE made a determination of exceptional circumstance under which applicants for funding from two DOE organizations—the Advanced Research Projects Agency-Energy and the Office of Energy Efficiency and Renewable Energy—may be required to include a U.S. Manufacturing Plan as part of their proposal. However, DOE patent counsel stated that it is too soon to determine how this action will affect the domestic

manufacture of inventions funded by those DOE organizations because development, patent, and commercialization time frames can each span years.

[37] The SunShot Initiative's mission is to make solar energy fully cost-competitive with traditional energy sources and supports efforts by private companies, universities, and national laboratories to reduce solar technology and grid integration costs, and accelerate solar deployment nationwide.

[38] The proposed domestic manufacturing provision stipulated that (1) at least 51 percent of the fair market value of each individual product embodying an invention or produced through the use of an elected invention shall be manufactured in the United States or (2) at least 51 percent of the all products embodying an invention or produced through the use of an invention shall be wholly manufactured in the United States.

[39] The proposed regulation explicitly states that all contractors must comply with all applicable U.S. laws regarding export control.

[40] The proposed regulation defines "change of control" as including: (1) any event by which any individual or entity other than the recipient becomes the beneficial owner of more than 50 percent of the total voting power of the voting stock of the recipient; (2) the recipient merges with or into any entity other than in a transaction in which the shares of the recipient's voting stock are converted into a majority of the voting stock of the surviving entity; (3) the sale, lease or transfer of all or substantially all of the assets of the recipient to any individual or entity other than the recipient in one or a series of related transactions; (4) the adoption of a plan relating to the liquidation or dissolution of the recipient; or (5) where the recipient is a wholly owned subsidiary at the time of award and the recipient's parent entity undergoes a change of control.

[41] According to DOE's proposed regulation, the $10 million threshold was selected so that it would apply to demonstration and large research and development projects but not smaller-scale financial assistance awards.

[42] The interagency Committee on Foreign Investment in the United States (CFIUS) also reviews transactions that could result in control of a U.S. business by a foreign person to determine their effect on the national security of the United States. If CFIUS finds that a covered transaction presents national security risks and that other provisions of law do not provide adequate authority to address the risks, then CFIUS may enter into an agreement with, or impose conditions on, parties to mitigate such risks or may refer the case to the President for action.

In: Department of Energy Funded Inventions ISBN: 978-1-63483-259-5
Editor: Peter Reed

Chapter 2

FEDERAL RESEARCH: INFORMATION ON THE GOVERNMENT'S RIGHT TO ASSERT OWNERSHIP CONTROL OVER FEDERALLY FUNDED INVENTIONS*

United States Government Accountability Office

ABBREVIATIONS

AIDS	Acquired Immune Deficiency Syndrome
DOD	Department of Defense
DOE	Department of Energy
GAO	Government Accountability Office
HIV	Human Immunodeficiency Virus
NASA	National Aeronautics and Space Administration
NSF	National Science Foundation
NIH	National Institutes of Health
NIST	National Institute of Standards and Technology

* This is an edited, reformatted and augmented version of The United States Government Accountability Office publication, GAO-09-742, dated July 2009.

Why GAO Did This Study

The Bayh-Dole Act, passed in 1980, allows recipients of federal research funds the option to retain patents on any inventions they create using those funds. At the same time, the act provides the government with rights intended to ensure that the public benefits from these federal research investments. One of these rights is known as the "march-in" authority, which allows federal agencies to take control of a patent when they have credible information that certain conditions described in the act have been met.

Until March 2009, the Bayh-Dole Act required GAO to report periodically on its implementation. To meet that requirement, for select federal agencies, GAO reviewed (1) the policies and procedures used to determine whether march-in authority should be exercised; (2) how the march-in authority has been used; and (3) what barriers and disincentives have been encountered in exercising the march-in authority.

GAO selected four agencies for this review that accounted for 89 percent of the federal research funding for fiscal year 2006. These were the Departments of Defense and Energy (DOD and DOE), the National Aeronautics and Space Administration (NASA), and the National Institutes of Health (NIH).

GAO is not making any recommendations in this report. DOE, NASA, and NIH provided technical comments on this report that GAO incorporated, as appropriate.

What GAO Found

Officials at DOD, DOE, NASA, and NIH rely on Commerce regulations for the Bayh-Dole Act and on their agencies' interpretations of the act to determine whether to exercise their march-in authority. Agency officials said that the administrative processes developed by Commerce are detailed and time-consuming, and may make it difficult to initiate and exercise a march-in proceeding. However, some officials said the detailed regulations ensure that appropriate and fair processes are followed during march-in proceedings. The agencies have chosen not to develop agency-specific guidance for a variety of reasons. For example, none of the officials believe that the regulations are onerous enough to warrant the development of agency-specific guidance and agency-specific guidance would reduce the flexibility agencies have to examine the specific circumstances of each case. In addition, an array of agency-specific regulations could hinder the transfer of research results to the market by increasing the regulatory burden on recipients of federal research funds.

None of the four agencies has chosen to exercise march-in authority. DOD, DOE, and NASA have neither discovered nor received information that would lead them to initiate a march-in proceeding or exercise their march-in authority during the last 20 years. In contrast, NIH has been petitioned formally three times, but in each case determined that the statutory requirements for march-in proceedings had not been met. Nevertheless, officials at DOD, NASA, and NIH said they value the authority because it provides leverage to promote commercialization of federally funded inventions. DOE officials disagree, in part, because no agency has ever exercised the authority. Agency officials said they do not have ongoing efforts to identify potential candidates for a march-in proceeding and primarily rely on the public, including potential competitors, to provide information that could lead to a march-in proceeding. According to these officials, their agencies would have to expend significant additional resources to track federally funded inventions because of the large number of inventions and because commercialization can take many years. Officials at DOD, NASA, and NIH said they value the march-in authority because it helps ensure that federally sponsored research results are commercialized. Also march-in authority is not the only tool to achieve the goals of the Bayh-Dole Act. For example, the government can take a patent without a license subject to reasonable compensation being paid to the patent owner or licensee that may allow for more timely interventions than would occur under the Bayh-Dole march-in process.

Federal and technology transfer officials identified four disincentives to the use of march-in authority. One of these is that the use of the march-in authority could have a “chilling effect” on federal research. These officials said that if a march-in occurred, investors would be less likely to provide the funds to commercialize federal inventions for fear of losing their investments. Also, because the march-in process can be long, these officials believe that it would have limited utility in an emergency situation. For example, the time to complete the fact-finding process in the three cases NIH reviewed ranged from 5 to 8 months.

July 27, 2009

The Honorable Patrick J. Leahy
Chairman

The Honorable Jeff Sessions
Ranking Member
Committee on the Judiciary United States Senate

The Honorable John Conyers, Jr.
Chairman

The Honorable Lamar Smith
Ranking Member
Committee on the Judiciary House of Representatives

Technological innovation is widely seen as responsible for much of the economic growth and increased standard of living in modern societies. Patent rights give inventors, or other patent owners, exclusive control over the use of their inventions for about 20 years, which promotes commercialization of new ideas and allows inventors to profit from their ideas. Patent rights ownership encourages the additional, and often substantial, investment of time and money needed to transform the technological innovations developed in the laboratory into goods, services, and processes available in the marketplace. Patent owners—including individuals, companies, and universities—may grant licenses to one or more businesses to complete this transformation and, in return, receive payments in the form of license fees or royalties.

The federal government supports technological innovation through a wide range of research activities that focus on the mission needs of various departments and agencies. In addition, it supports work in areas where a specific need has been identified that the private sector has not addressed. Although the largest share of research funding comes from the private sector, the federal government funds a majority of the nation's basic research, which produces the innovations that drive technological progress. Moreover, federal support accounts for over half of the research conducted at colleges and universities in the United States. Because the public benefits when technological advances are transformed into new goods and services in the marketplace, the federal government has an interest in facilitating the commercialization of new inventions that arise from the research that it funds.

Since its enactment in 1980, the Bayh-Dole Act has provided recipients of federal research and development funding—often referred to as contractors—the option to retain patents on the inventions they create, provided they adhere to certain requirements.[1] A main goal of the act is to promote the utilization of inventions arising from federally supported research or development, and observers have judged the act a success in this regard. Prior to 1980, when the government routinely retained the patents on federally sponsored inventions, only 5 percent of these patents were ever used in the private sector. In contrast, some stakeholders, including federal and technology transfer officials, today believe inventions that arise from federally funded research are routinely commercialized, although comprehensive data are not available on how often this happens. Each federal agency that enters into funding agreements subject to the Bayh-Dole Act is responsible for administering the act's requirements and the implementing regulations developed by the Department of Commerce.

The Bayh-Dole Act also provides the federal government with certain rights to protect the public against nonuse or unreasonable use of federally funded inventions. One of these rights, known as the "march-in" authority, authorizes federal agencies, at their discretion, to require the contractor or licensee to grant a license to any responsible entity or entities when credible information exists that certain statutory conditions in the act have been met. For example, an agency may march in if it determines that an inventor is not taking the necessary steps toward commercialization of the technology, or that such action is needed to meet public health or safety needs. [2]

Until recently, the Bayh-Dole Act also contained a requirement that GAO issue a report on how agencies have implemented the act's provisions at least once every 5 years. In consultation with your offices we began work on this review to meet that reporting requirement. However, subsequent to our initiating this review, the Omnibus Appropriations Act for fiscal year 2009, eliminated the recurring study requirement on March 11, 2009.[3] As agreed with your offices, we have completed this review and addressed the following objectives: (1) what policies and procedures have federal agencies with significant research budgets established to determine whether march-in authority under the Bayh-Dole Act should be exercised; (2) to what extent have these selected federal research agencies used Bayh-Dole march-in authority and what do they believe are its benefits; and (3) what barriers and disincentives, if any, have these agencies encountered to the exercise of their march-in authority under the BayhDole Act.

To determine which agencies to focus our review on, we analyzed federal research and development budgets for all federal research agencies. We selected the Department of Defense (DOD), the Department of Energy (DOE), the National Aeronautics and Space Administration (NASA), and the National Institutes of Health (NIH) within the Department of Health and Human Services because together they accounted for 89 percent of the total federal research funding for fiscal year 2006—the most recent year for which complete data were available. For each of the objectives, we reviewed key agency documents and interviewed officials from the technology transfer offices of each agency. In addition, for each of the objectives we spoke with officials in stakeholder groups such as the Association of University Technology Managers, the Biotechnology Industry Organization, and the American Intellectual Property Law Association, as well as academics who have evaluated the Bayh-Dole Act.

We conducted our work from November 2008 to July 2009 in accordance with generally accepted government auditing standards. Those standards require that we plan and perform the audit to obtain sufficient, appropriate evidence to provide a reasonable basis for our findings and conclusions based on our audit objectives. We believe that the evidence obtained provides a reasonable basis for our findings and conclusions based on our audit objectives.

BACKGROUND

The Bayh-Dole Act was enacted in 1980, in part, to address the low utilization rate of federal patents. At the time the bill was considered, 26 different federal agency policies existed regarding the use of results from federally funded research. Prior to Bayh-Dole's enactment, agencies frequently retained title to inventions made with federal support whether the research was performed in federal laboratories, in universities, or by individual companies. Licenses to use and further commercialize the patents on federally funded inventions were then negotiated with firms typically on a non-exclusive basis or, more rarely, for the exclusive use by one manufacturer. The Bayh-Dole Act established a governmentwide policy that gave contractors the opportunity to retain ownership of federally funded inventions. In addition, it was designed to use the patent system to promote the utilization of inventions arising from federally supported research or development and to encourage maximum participation of small business firms in federally supported research and development efforts, among other things. Many experts continue to believe that certainty in the ownership of patents and exclusivity in the right to develop the related technology are important for both large and small firms.[4]

In exchange for the right to retain ownership of federally sponsored inventions under the Bayh-Dole Act, contractors must agree to certain reporting requirements. More specifically, contractors agree to notify the funding agency within 2 months after the contractor learns that an invention has been created and to notify the funding agency within 2 years after this notification of the contractor's decision to retain title to the invention. In addition, contractors agree to apply for a patent on the invention typically within 1 year of the election of title, attempt to commercialize the invention, and to provide additional reports. These additional reports, if requested by the agency, can provide such information as utilization of the invention and patent-related information such as the filing date, patent application number and title, and patent number and issue date for the invention in any country in which the contractor has applied for a patent. Failure by the contractor to disclose the invention, elect title to it, or file a patent application within the times specified, or failure to follow through with the patent application process, allows the relevant federal agency to obtain ownership of the invention.

The Bayh-Dole Act also reserved certain rights for the government to protect the public's interests. Specifically, the government retains "a nonexclusive, nontransferable, irrevocable, paid-up license to practice or have practiced for or on behalf of the United States any subject invention throughout the world," also known as a nonexclusive royalty-free license. In addition, the act provides the government march-in authority. Under this authority, the federal agency that funded the development of an invention has the right to require the contractor or exclusive licensee to grant a license in any field of use to a responsible applicant upon terms that are reasonable under the circumstances, if the agency determines that:

- the contractor has not made, and is not expected to make, efforts to commercialize the invention within an agreed upon time frame;
- public health or safety needs are not reasonably satisfied by the contractor or licensee;
- the use of the invention is required by the federal government and the contractor or licensee cannot meet the government's requirements; or

- the owner of an exclusive license is not ensuring that the invention is "manufactured substantially" in the United States and has not obtained the necessary waivers to do so.

Implementation of the Bayh-Dole Act is decentralized across the federal government. Each federal agency that enters into funding agreements subject to the Bayh-Dole Act is responsible for administering the act's requirements. However, the act directs the Department of Commerce to develop regulations to implement the provisions contained in the act, including procedures for agencies to follow regarding the exercise of the march-in authority.[5] The regulations Commerce issued in 1987 also allow agencies to develop supplemental procedures regarding their march-in authority. Although Commerce does not maintain any overall Bayh-Dole databases, other agencies rely on Commerce as a coordinator and consultant for Bayh-Dole related issues.

The regulations established by Commerce detail the procedures an agency must follow when it receives information that it believes might warrant the exercise of march-in rights. Specifically, the agency must notify the contractor, in writing, that it has information it believes might warrant the exercise of its march-in authority. As part of this notification, the agency is to provide the contractor 30 days to respond informally, either verbally or in writing, with relevant information. Once the agency has received the contractor's response, it may initiate the march-in procedures within 60 days through written notice to the contractor and its assignee or exclusive licensee, as appropriate and if known to the agency. The notice must include the reasons for the proposed march-in and the specific uses of the invention for which the agency may require licensing. Within 30 days after receiving written notice of the proposed march-in proceeding, the contractor may submit information opposing the proposed march-in to the agency in person, in writing, or through a representative. If the agency determines that the contractor's information raises a dispute over the facts of the case, it must undertake a fact-finding process that gives the contractor the opportunity to appear with counsel, submit documents, present witnesses, and question individuals presented by the agency. The results of the fact-finding process and a recommendation are presented to the head of the agency (or his or her designee) as well as to the contractor. Both the agency and the contractor have 30 days to submit written arguments to the head of the agency or designee. In addition, the contractor may request to present oral arguments. Within 90 days after the completion of the fact-finding or oral arguments, whichever is later, the agency must provide a written decision regarding whether march-in rights will be exercised. Any decision unfavorable to the contractor will be held in abeyance pending the exhaustion of the contractor's administrative and judicial appeals. At any point, the agency may terminate the fact-finding process if it decides not to exercise its march-in authority.

The time from when an agency announces a funding opportunity to the time a viable commercial product reaches the marketplace may take many years and substantial financial investment. During the period of agency funding, which may last 8 to 10 years for drugs and biologics, the agency's program, procurement, and/or grants office monitors the progress of the research and maintains contact with the contractor. In fiscal year 2007 federal agencies devoted $116 billion to conduct research on various topics related to their respective missions. Pharmaceutical-related inventions, which may arise from research sponsored by NIH, may require an additional 10 to 15 years after the invention is made to obtain the federal approvals necessary to reach the market. According to industry officials, pharmaceutical-related inventions may require an investment of between $800 million and $1.3 billion to conduct the safety and other studies required for approval.

Additional time may be required to obtain a patent on the invention and to develop a market ready process or product. [6] More specifically, the U.S. Patent and Trademark Office issues a patent in 32 months, on average, but the time ranges from 28 months for inventions in the fields of semiconductors and electrical items to almost 44 months for computer software and communications inventions. Once a patent is granted, the patent owner has, in most instances, a period of 20 years from the date the application was filed during which time the patent owner has the right to exclude others from making, using, or selling the patented invention.

Federal Agencies Use Department of Commerce Regulations to Implement the March-in Authority under the Bayh-Dole Act

Officials at DOD, DOE, NASA, and NIH rely on Commerce regulations for the Bayh-Dole Act and on their agencies' interpretations of the act to determine whether to exercise their march-in authority. These officials told us that the administrative processes developed by Commerce for agencies to use when considering whether marching-in may be warranted are detailed and time-consuming, and may make it difficult to initiate a march-in proceeding. However, some officials also acknowledged that because the regulations are detailed, they ensure that appropriate and fair processes are followed during march-in proceedings. One official noted that there is no way to pre-empt the process and retain the necessary legal protections for all of the participants in the process. According to this official, the regulations, while detailed and time-consuming, allow everyone to be heard during the process. For example, during the fact-finding procedure the contractor has the opportunity to appear with counsel, submit documentary evidence, present witnesses, and cross-examine witnesses who the agency presents. Moreover, both the contractor and agency staff have an opportunity to rebut an agency's decision and contractors may appeal adverse decisions to the federal courts, which delays action on the agency's decision until the appeals process is concluded.

However, according to agency officials we spoke with, the agencies have chosen not to develop agency-specific guidance for a variety of reasons. First, none of the agency officials we spoke with believe that the regulations developed by Commerce are onerous enough to warrant the development of agency-specific guidance. Second, both agency and technology transfer officials told us that agency-specific guidance would, in essence, pre-define how the federal government would exercise its march-in authority and reduce the flexibility agencies have to examine the specific circumstances of each individual case. Third, federal officials—as well as officials from organizations that represent technology transfer offices in colleges and universities—told us that creating an array of agency-specific regulations could hinder the transfer of research results to the market by increasing the regulatory burden on contractors.[7] For example, one technology transfer official said that many universities receive funding concurrently from more than one federal agency. In such cases, these contractors could be required to follow a different set of regulations from each of their agency partners. As a result, these officials believe that Commerce should remain in charge of developing march-in regulations, rather than have individual agencies issue their own policies and procedures. Finally, technology transfer officials we spoke to also said that march-in regulations should be centralized at a high enough level to ensure consistency among federal research agencies in their march-in decisions.

Until August 2007, if federal agencies or contractors had any questions concerning Bayh-Dole Act implementation issues, including march–in procedures, they generally coordinated with officials in Commerce's Technology Administration. However, since August 2007, as a result of changes mandated by the America COMPETES Act, the Technology Administration has been disbanded and Commerce has shifted responsibility for the Bayh-Dole Act to the National Institute of Standards and Technology (NIST). Officials from two technology transfer organizations told us that, as a result of this change, the department currently has little expertise on the march-in process. Specifically, technology transfer officials told us they were concerned that NIST did not have the knowledge and experience of the Technology Administration with regard to oversight of march-in procedures and officials at one organization believed that this might cause some ambiguity in facilitating agencies' implementation of the act.

NIST officials acknowledged that no one currently in their office has any experience with the march-in authority and said the process appears to be very time-consuming and complex. However, these officials told us that when the Technology Administration was disbanded, the same lawyers who worked on Bayh-Dole issues continued to provide their services, which allowed continuity in the overall legal aspects of oversight for the act. They also noted that most of the questions they have addressed for agencies concern aspects of the act other than the march-in authority. They also believe that because agencies are not required to contact NIST with questions related to the Bayh-Dole Act, that NIST's role in any future march-in proceedings will likely be very limited.

None of the Agencies We Reviewed Has Used March-in Authority, but Three Value It As a Way to Promote Commercialization of Inventions

None of the four agencies we reviewed has chosen to exercise march-in authority under the Bayh-Dole Act. DOD, DOE, and NASA have neither discovered nor received information that would lead them to initiate a march-in proceeding or exercise their march-in authority during the last 20 years. In contrast, NIH has been petitioned formally to exercise its march-in authority three times, but in each case determined that the statutory requirements for march-in proceedings had not been met. Nevertheless, officials at three of the four agencies told us they value the authority because, together with other tools, it provides them leverage to promote commercialization of federally funded inventions. In contrast, DOE officials do not believe march-in authority has significant value as leverage, in part, because no agency has ever exercised the authority.

Officials at all four agencies included in our review acknowledged that their agencies have not conducted any march-in proceedings. They further acknowledged that while they monitor contractors' compliance with reporting requirements, their agencies do not have ongoing efforts to identify potential candidates for march-in proceedings from the wide-array of federally funded inventions. These officials told us that they primarily rely on public and private sources of information, including news reports, interest groups, and potential competitors, to provide them with information that could lead to a march-in proceeding. For example, according to one official, participants in the science and technology market are very aware of emerging technologies and information on patents is publicly available, which allows interested entities to know what inventions and technologies are being developed. In addition, companies employ technology scouts to report on the technologies being produced by other companies. Officials told us that one source of information regarding a potential march-in proceeding could be a person or business that wants to enter into a licensing agreement but is unable to negotiate agreeable terms. However, they also acknowledged that such instances are generally uncommon because most contractors are very interested in licensing their inventions.

According to the agency officials we spoke with, relying on the public for information is a more efficient and effective mechanism for tracking federally funded inventions, which would otherwise require federal agencies to expend significant additional resources to monitor a large volume of federally funded inventions for possible situations that might lead to march-in proceedings. In fiscal year 2008, NIH provided 50,980 awards, worth about $21 billion, to 2,606 institutions. The agency's awards for the previous 5 fiscal years were steady at about this same level. Monitoring such a large number of awards and institutions would be very resource intense. Moreover, because many inventions require substantial investments of time to produce a market-ready product or process, agencies would need to monitor awards and their subsequent inventions over a number of years. For example, NIH officials said that pharmaceutical inventions may take as many as 14 years to reach the marketplace. In addition, although contractors report information on inventions that result from federally funded projects, they are not required to report information on progress toward commercialization of those inventions or other details of the licensing agreements they enter into, which are considered proprietary information. Consequently, agencies do not always receive information on the extent to which licensees are making progress toward commercialization of the inventions the agencies have funded. Officials also told us that proactive efforts to track federally funded inventions are further complicated by the fact that a single invention may result in multiple licensing agreements for different uses. For example, a contractor who owns a cancer treatment could license the technology to one entity to treat eye cancer and to another to treat liver cancer.

Since Congress enacted the Bayh-Dole Act in 1980, only NIH, of the four agencies we reviewed, has received formal march-in petitions—one in 1997 and two in 2004. In each of these cases, the agency determined after a 5- to 8-month fact-finding process that the circumstances did not meet any of the statutory conditions under which march in could occur. Specifically, in 1997, NIH received a petition in which the petitioner alleged that the invention's owner and exclusive licensee had failed to take reasonable measures to bring a stem cell separation device to market and that doing so would alleviate patient health and safety needs. NIH found no basis to initiate a march-in proceeding because it determined that the invention's owner and exclusive licensee had taken effective steps to develop the device and that it was already being marketed. In 2004, NIH received two more petitions, in which the petitioner expressed concern that the price of two drugs—one to treat HIV/AIDS and the other to treat glaucoma—made them unaffordable for many people living with these diseases, posing a threat to their health and safety. However, NIH determined that the drugs were already on the market and widely prescribed, and therefore marching in would not alleviate health and safety needs that were not already being satisfied by the producer. NIH also stated in its decisions that drug pricing is an issue more appropriately left to the Congress.[8] Furthermore, as NIH noted in its decision on the 1997 petition, the agency is "wary of forced attempts to influence the marketplace for the benefit of a single company."

Although DOE has not been petitioned to exercise its march-in authority nor has it used the authority on its own, the department used the BayhDole march-in framework to review a dispute brought by a company against a contractor and its exclusive licensee over the use of two inventions that could identify gene sequences. According to the company, the contractor and its licensee had not taken effective steps to achieve substantial utilization of the inventions, and had not given the requisite preference to small businesses. While this dispute did not arise under the Bayh-Dole Act, DOE suggested, and all parties agreed, to settle the dispute using the march-in procedures detailed in the Commerce regulations. During a 30-month fact-finding process, both parties to the dispute submitted evidence and counter evidence and reviewed the draft decision prior to its release. DOE decided not to march in based on its determination, among other things, that the terms of the exclusive license were fair and that the company making the allegations had failed to offer sufficient evidence to support its contentions.

Although none of the agencies we reviewed has actually used its authority to march in, officials in three of the four agencies we contacted said they value the authority and they do not want it eliminated because it helps to ensure that federally sponsored research results are commercialized. These officials told us that the march-in authority is particularly valuable as leverage in informal discussions between contractors and sponsoring agencies and in license negotiations between contractors and potential licensees to encourage commercialization of technologies developed with federal funding. However, neither the agencies we reviewed nor the technology transfer organizations we contacted maintain data on the extent to which the potential for a march-in proceeding is discussed informally during negotiations.

According to some agency and technology transfer officials, the parties to licensing negotiations are usually sufficiently aware of the potential for march-in that it may not be necessary to explicitly discuss this possibility during meetings. However, neither could provide us with any metrics by which they could measure this effect, and no data exist on the extent to which contractors or licensees are aware of the potential for an agency to march in and to what extent this influences their decisions. Executives from two bio-technology firms told us that they are well aware of the Bayh-Dole Act and its march-in provision. They consider the potential for a march-in as one of several business risks, but said it is not a subject they typically discuss during licensing negotiations. Nevertheless, according to one technology transfer official, an explicit discussion of march-in authority can provide effective leverage to push a company struggling to meet its obligation to pursue commercialization of a federally funded invention. DOE officials, on the other hand, said an awareness of the march-in authority did not appear to have much influence on its contractors and their licensees. DOE officials said their contractors generally produce inventions, processes, and technologies that are intended for the market and are already strongly motivated by potential profits to move forward quickly. Consequently, these officials said it is difficult to see how the potential for a march-in proceeding under the Bayh-Dole Act would provide an additional incentive to these contractors.

Although most of the officials we spoke with value the leverage that the march-in authority provides, they said they prefer to work with contractors informally to resolve commercialization issues. For example, NIH officials noted that contractors often resolve such issues—without agency involvement—by reviewing the milestones in the licensing agreement to determine whether the licensee has met its obligations, and if it has not, the contractor may adjust the terms of the agreement based on speed and results of the licensee's efforts or revoke the license and seek a new licensee. According to one NIH official, if NIH enters such discussions its mere involvement often serves as enough leverage to encourage resolution of the problem without resorting to an explicit mention of march-in. Similarly, DOD officials said that in the early 1990s, during a patent-related dispute between two defense contractors, one of the companies raised the possibility of petitioning DOD for a march-in proceeding to settle the disagreement. DOD entered into informal discussions with both companies, then withdrew, and the companies subsequently resolved their dispute without petitioning for a march-in.

Some officials also told us that the march-in authority is not the only available tool to achieve commercialization goals for federal research efforts or to meet the government's needs. For example, NIH officials told us that one useful tool is the agency's guidance for contractors to use when negotiating license agreements. Contractors can enter into such agreements with parties who wish to commercialize an invention. The NIH guidance recommends including specific commercialization milestones and a termination clause to ensure that inventions are commercialized by licensees. For example, if an invention has a potential therapeutic use, the agreement may include requirements to reach federal approval for various clinical trials by certain dates, as well as the anticipated date of first sale. Technology transfer officials we spoke with said that the widespread use of commercialization milestones and termination clauses reduces the likelihood that an agency would need to march in because contractors are already assuring that commercialization is achieved.

The government can also use patented technology without a license subject to reasonable compensation being paid to the patent owner or licensee, regardless of whether the invention had been developed with federal funding. This allows the federal government to use an invention without a license, if the use is "by or for the United States." Further, under federal law if the federal government uses a patent, the patent owner or licensee may sue the government to recover reasonable compensation but may not stop the government from using the invention.[9] This option might be of greater value than the Bayh-Dole march-in authority in the case of a public health emergency because it allows for rapid action and it allows the government to use inventions that incorporate federally funded technologies as well as technologies that were not federally funded. In addition, officials told us that the Bayh-Dole Act itself contains another tool—the royalty-free license—that allows federal agencies to use federally funded inventions without risk of infringing the ownership rights of the contractor or licensee. For example, federal agencies may contract with a third party to manufacture products containing such inventions for, or on behalf of, the government. However, if the product or process contained inventions that were not developed with federal funds, the government would need to negotiate a license to use them. Finally, some agencies, including DOD, DOE, and NASA, have been granted other statutory tools that provide additional flexibility to negotiate ownership terms with contractors. For example, all three have similar statutory authorities—called "other transaction authority"—that apply to certain research efforts conducted under contracts. DOD and DOE have used this authority to obtain cutting-edge research and prototypes for their use and NASA has used its authority to negotiate ownership rights that will foster the commercialization of inventive work produced under collaborative research projects that are not being conducted specifically for the agency.

FOUR KEY CONCERNS MAY CREATE DISINCENTIVES TO THE USE OF BAYH-DOLE MARCH-IN AUTHORITY BY FEDERAL AGENCIES

Four key disincentives inhibit federal agencies use of Bayh-Dole march-in authority. First, the potential "chilling effect" that such an action might have could deter investors from investing in the commercialization of the research results and some researchers from participating in federal research efforts. Second, the lengthy march-in process could be unworkable in an emergency or other time-critical situation. Third, commercial products or processes based on federal inventions sometimes employ multiple patents, some of which are not federally funded. Such circumstances often pose difficult, if not intractable, issues that could make marching in unattractive for federal officials seeking to commercialize an invention. Finally, agencies might be disinclined to march in if current licensees have specialized knowledge that makes them particularly well positioned to bring a product to market, and if the loss of such knowledge through a march-in proceeding might jeopardize the commercialization of an invention. This section further describes these four disincentives.

Some agency, university, and industry officials we contacted said the march-in authority could have a "chilling effect" on the willingness of venture capital firms and other investors to provide funding for the further commercial development of federally funded inventions. For example, three of the technology transfer officials we contacted said the chilling effect on investors would be increased if agencies used the march-in authority under circumstances that were not well supported by the facts. According to these officials, investors are looking for profitable technologies and inventions that either have, or are close to obtaining, a patent, which allows them to capture profits in relative safety. They said that for some investors the mere existence of an agency's march-in authority makes such investments more risky because, should an agency actually exercise its authority, investors may believe the value of their investment could evaporate or decline significantly and these perceived risks could increase significantly.

However, executives from two bio-technology firms—both of which hold licenses to commercialize technologies developed in part with federal funds and which must raise money from investors to pursue commercialization—told us the perceived risk that an agency might march in is far less important to investors than other risks they face. For example, they cited the product's likely efficacy as perhaps the key factor for investors to consider in making such decisions. The executives added that one of the greatest concerns of their potential investors is how soon the product or process can be marketed and, as a result, return a profit on their investment. These executives expressed confidence that if licensees take care to follow the requirements of the Bayh-Dole Act, then march-ins would be rare and should not negatively affect the flow of federally funded inventions to the market. In addition, technology transfer officials noted that at the time the act was passed companies were often unwilling to enter into licensing agreements due to concerns about how agencies would use the authority. They said such concerns have diminished, in part because the small number of fact-finding proceedings has not led agencies to march in.

The march-in authority would also have a chilling effect if researchers, particularly private-sector researchers, were unwilling to apply for federally funded projects because the potential for an agency to march in creates uncertainty with regard to ownership of an invention. However, none of the officials we contacted was aware of specific instances when a researcher had declined to apply for federal funding and they said it is impossible to know the extent to which researchers decide against applying for federal funds due to such concerns. In contrast, officials at DOE said they do not believe the potential for a march in is a concern for their contractors. For example, DOE officials noted that following the release of a recent solicitation, 60 small businesses called with questions about the march-in authority. However, even after DOE officials explained the march-in authority to these callers, overwhelmingly they submitted applications.

Because the march-in process itself can be long and the outcome unknown pending a possible appeal of the agency's decision to the federal courts, the NIH officials we contacted believe march-in authority could have limited utility in an emergency situation, such as an important public health issue, that required prompt federal action. More specifically, the Commerce regulations that govern march-in procedures provide for a quasi-judicial process that may require more time to complete than the other legislative options mentioned above. The march-in procedures allow for contractors to be represented by counsel, the opportunity to call and confront witnesses, and the chance to introduce documentary evidence and review the evidence others have presented. In the four fact-finding instances we reviewed, the time to reach a decision not to initiate march-in proceedings ranged from 5 to 30 months. According to NIH officials, the specifics of each Bayh-Dole fact-finding effort are likely to vary, but the process to determine whether a march-in proceeding is warranted will usually require at least several months to accomplish. Moreover, in the event an agency decides to march in, action on the decision may be delayed pending review by the federal courts if the contractor or licensee appeals the decision. In emergency situations, NIH officials said the government could use other legal authorities, discussed above, to obtain the necessary rights.

Officials at NASA and NIH also reported that a march-in proceeding would be complicated by the fact that most products and processes include multiple technologies covered by multiple patents, and that in many cases only some of them have been developed with federal funding. As a result, federal agencies may only have the authority to march in on one aspect of a product or process, yet marching in may negatively affect the value of all the other patented inventions associated with the product or process. For example, NIH officials described the development of a single genome test that used 17 patents from 13 organizations (3 from outside the U.S.), some of which used government funding and some did not. These officials said it would be impossible for NIH to determine that 1 of those 17 patents is not being commercialized fast enough, or not meeting a health need, in the face of its dependence on 16 other patents. Any such effort would require the cooperation of 12 other organizations, and an unknown number of licensees. The officials concluded that it would be an impossible task for NIH, or any agency, to decide to march in under those circumstances.

Officials at NIH also said that agencies might be disinclined to march in if current licensees have specialized knowledge about how to bring a particular product to market. If the loss of such knowledge would jeopardize the commercialization of an invention, agencies might be reluctant to pursue a march-in. For example, licensees may possess information such as trade secrets, other patented technologies related to product development, experience with the federal approval process, or marketing experience. If NIH were to force a contractor or licensee to grant a license to another entity, it would have to consider whether the other patented technologies would be available to the new licensee and whether the new licensee would have the knowledge, resources, and commitment needed to commercialize the product.

AGENCY COMMENTS AND OUR EVALUATION

We provided a copy of a draft of this report to DOD, DOE, NIH, and NASA for their review and comment. In commenting on the draft, NASA stated that the report provides a balanced view of the issues related to the regulations associated with the Bayh-Dole Act. NASA also provided technical comments that we incorporated, as appropriate. DOD, DOE, and NIH did not provide overall comments, but NIH provided technical comments that we incorporated, as appropriate.

Anu K. Mittal
Director, Natural Resources and Environment

APPENDIX I: OBJECTIVES, SCOPE, AND METHODOLOGY

The objectives of this report were to determine (1) what policies and procedures federal agencies with significant research budgets have established to determine whether march-in authority under the Bayh-Dole Act should be exercised; (2) the extent to which these selected federal research agencies have used Bayh-Dole march-in authority and what they believe are its benefits; and (3) what barriers and disincentives, if any, these agencies have encountered to the exercise of their march-in authority under the Bayh-Dole Act.

We sought to focus our review on those federal agencies whose combined research and development spending represent a significant portion of total federal research and development spending. To identify the federal research agencies that meet this criterion, we obtained and analyzed research and development funding data from the National Science Foundation (NSF) on preliminary federal obligations for research and development for all federal research agencies. The top four agencies receiving research and development funding were the Department of Defense (DOD), the Department of Energy (DOE), the National Aeronautics and Space Administration (NASA), and the National Institutes of Health (NIH) within the Department of Health and Human Services. We judgmentally selected these four agencies as the focus of our review. We compared the combined percent of funding from the total research and development allotment for these four agencies to the total allotment for the federal government and found that these four agencies accounted for approximately 89 percent of the total federal research funding for fiscal year 2006—the most recent year for which NSF had complete data. In assessing the reliability of the NSF data, we noted that it reports a 100 percent response rate, with responses to all items; thus, we determined it was sufficient for the purposes of this analysis.

To gain insights into the history of the Bayh-Dole Act, including its provision for march-in authority, as well as to understand the context in which the law was enacted and its current environment, we reviewed the act's legislative history, including congressional hearing statements made by the act's sponsors and other stakeholders. We also reviewed the available literature on the Bayh-Dole Act's implementation and the effects it has had on federal research. To understand the law's requirements, we reviewed all provisions of the act, giving special emphasis to those sections that establish march-in authority. To understand how agencies are to implement their responsibilities under the act, we reviewed the Department of Commerce's Bayh-Dole regulations.

For each of our three objectives, we interviewed officials from the technology transfer offices and offices of general counsel at DOD, DOE, NASA, and NIH, as well as officials from the National Institute of Standards and Technology. In addition, we contacted officials in stakeholder groups such as the Association of University Technology Managers, the Biotechnology Industry Organization, the American Intellectual Property Law Association, the Association of Public and Land-grant Universities, and Essential Inventions, as well as academics who have evaluated the Bayh-Dole Act. We also contacted representatives from the biotechnology industry who invest in and/or develop federally funded technologies. We reviewed the three march-in petitions that NIH received, and NIH's determinations in these cases, to understand how NIH applies the Commerce regulations. Finally, we studied NIH's research tool guidelines to determine their impact on agency decisions on whether to conduct march-in proceedings.

We conducted our work from November 2008 to July 2009 in accordance with generally accepted government auditing standards. Those standards require that we plan and perform the audit to obtain sufficient, appropriate evidence to provide a reasonable basis for our findings and conclusions based on our audit objectives. We believe that the evidence obtained provides a reasonable basis for our findings and conclusions based on our audit objectives.

End Notes

[1] The term "contractor" means any person, small business firm, or nonprofit organization that is a party to a federal funding agreement, which includes contracts, grants, or cooperative agreements for the performance of experimental, developmental, or research work.

[2] The two additional statutory conditions under which agencies may exercise march-in authority are (1) the use of an invention is required by the federal government and the contractor cannot meet the government's requirements; and (2) the patent owner or exclusive licensee has failed to take certain steps to ensure that any products embodying the invention or produced through the use of the invention will be manufactured substantially in the United States.

[3] Pub. L. No. 111-8, Div. G, Title I, section 1301(h), 123 Stat. 829 (2009).

[4] The Bayh-Dole Act by its terms applies to universities, non-profit organizations, and small businesses that receive federal research funding. A presidential memorandum in 1983, followed by an Executive Order in 1987, directed federal agencies, to the extent permitted by law, to establish policies for all businesses that are substantially the same as those contained in the Bayh-Dole Act.

[5] As originally enacted, the act required the Office of Federal Procurement Policy to develop these regulations. In 1984 Congress transferred this regulatory authority to the Department of Commerce. Pub. L. No. 98-620, § 501(10), 1984.

[6] The time required to obtain a patent may overlap with the period of federal funding for the research.

[7] Throughout this report we refer to officials from organizations that represent technology transfer offices in colleges and universities as technology transfer officials.

[8] In its decision on the drug for treating HIV/AIDS, NIH also stated that the Federal Trade Commission was the appropriate agency to address allegations that the drug manufacturer had engaged in anti-competitive practices.

[9] 28 U.S.C. § 1498(a).

In: Department of Energy Funded Inventions ISBN: 978-1-63483-259-5
Editor: Peter Reed

Chapter 3

THE BAYH-DOLE ACT: SELECTED ISSUES IN PATENT POLICY AND THE COMMERCIALIZATION OF TECHNOLOGY*

Wendy H. Schacht

SUMMARY

Congressional interest in facilitating U.S. technological innovation led to the passage of P.L. 96- 517, Amendments to the Patent and Trademark Act (commonly referred to as the Bayh-Dole Act after its two main sponsors). The act provides patent rights to certain inventions arising out of government-sponsored research and development (R&D) to non-profit institutions and small businesses with the expressed purpose of encouraging the commercialization of new technologies through cooperative ventures between and among the research community, small firms, and industry.

Patents provide an economic incentive for companies to pursue further development and commercialization. Studies indicate that research funding accounts for approximately one-quarter of the costs associated with bringing a new product to market. Patent ownership is seen as a way to encourage the additional, and often substantial investment necessary for generating new goods and services in the private sector. In an academic setting, the possession of title to inventions is

* This is an edited, reformatted and augmented version of Congressional Research Service, Publication No. RL32076, dated December 3, 2012.

expected to provide motivation for the university to license the technology to companies for commercialization in expectation of royalty payments.

The Bayh-Dole Act has been seen as particularly successful in meeting its objectives. However, while the legislation provides a general framework to promote expanded utilization of the results of federally funded research and development, questions have been raised as to the adequacy of current arrangements. Most agree that closer cooperation among industry, government, and academia can augment funding sources (both in the private and public sectors), increase technology transfer, stimulate more innovation (beyond invention), lead to new products and processes, and expand markets. However, others point out that collaboration may provide increased opportunities for conflicts of interest, redirection of research, less openness in sharing of scientific discovery, and a greater emphasis on applied rather than basic research. Additional concerns have been expressed, particularly in relation to the pharmaceutical and biotechnology industries, that the government and the public are not receiving benefits commensurate with the federal contribution to the initial research and development.

Actual experience and cited studies suggest that companies which do not control the results of their investments—either through ownership of patent title, exclusive license, or pricing decisions—tend to be less likely to engage in related R&D. The importance of control over intellectual property is reinforced by the positive effect P.L. 96-517 has had on the emergence of new technologies and techniques generated by U.S. companies.

INTRODUCTION

Congressional interest in facilitating U.S. technological innovation led to the passage of P.L. 96- 517, Amendments to the Patent and Trademark Act, commonly referred to as the "Bayh-Dole Act" after its two main sponsors former Senators Robert Dole and Birch Bayh. Under this 1980 law, as amended, title to inventions made with government support may be provided to the contractor if that contractor is a small business, a university, or other non-profit institution. The legislation is intended to use patent ownership as an incentive for private sector development and commercialization of federally funded research and development (R&D). As a response to congressional efforts to create a unified government patent policy pertaining to inventions made with federal support, the Bayh-Dole Act promotes cooperative activities among academia, small business, and industry leading to new products and processes for the marketplace.

This report discusses the rationale behind the passage of P.L. 96-517, its provisions, and implementation of the law. Observers generally agree that the Bayh-Dole Act has successfully met its objectives. However, some experts argue that the issues associated with the law's patent policies should be revisited given the current R&D environment. Much of the renewed interest is a result of the legislation's effect on the biotechnology and pharmaceutical industries where critics assert that the private sector is receiving benefits to the detriment of the public interest. Other analysts, particularly in the defense arena, maintain that the existing rights maintained by the government are too restrictive and prevent industry from meeting national needs. Many of these issues and concerns are similar, if not identical to those addressed during the 15 to 20 years of deliberation prior to enactment of the law. These too will be explored to provide a context for current discussions.

An Historical Perspective

The Rationale

In the late 1970s, the United States Congress was involved in a series of legislative debates over ways to promote private sector development and utilization of federally funded research and development. This was soon followed by expanded congressional interest in additional means to foster technological advancement and commercialization in industry. During the 1980s and 1990s, various initiatives resulted in laws designed to encourage increased innovation-related activities in the business community and to remove barriers to technology development, thereby permitting market forces to operate.[1] Laws promoting cooperative R&D and/or joint ventures involving the federal government, industry, and academia have been a cornerstone of the majority of these efforts and include legislation that created a system to transfer technology from federal laboratories to the private sector; implemented tax incentives for collaborative work; instituted direct and indirect government support for increased R&D; and changed government patent policy to provide an economic inducement for commercialization of federally funded technology, the subject of this report.

P.L. 96-517, the Bayh-Dole Act, was one of the first of these initiatives. Prior to 1980, only 5% of government owned patents were ever used in the private sector although a portion of the intellectual property portfolio had potential for further development, application, and marketing. The Bayh-Dole Act was constructed, in part, to address the low utilization rate of these federal patents. The House report to accompany H.R. 6933 (the House counterpart to the Senate bill that eventually became the Bayh-Dole Act) noted that, at the time the bill was considered, 26 different agency policies existed regarding the use of the results of federally funded R&D. Generally the government retained title to inventions made with government support whether the research was performed in federal laboratories, in universities, or by individual companies. Licenses to use government patents were then negotiated with firms either on a non-exclusive basis (meaning additional companies could use the technology) or, more rarely, for the exclusive use by one manufacturer. However, it was widely argued that without title (or at least an exclusive license) to an invention and the protection it conveys, a company would not invest the additional, and often substantial time and money necessary to commercialize a product or process for the marketplace.

In 1980, the federal expenditure for research and development totaled $55.5 billion (in constant 2000 dollars).[2] The money typically was used to support research and development to meet the mission requirements of the federal departments and agencies (e.g., defense, public health, environmental quality) or to finance work in areas where there was an identified need for research, primarily basic research, not being performed in the private sector. While the government's investment led to many new inventions that have profoundly influenced our society, many in Congress were of the opinion that additional applications could be pursued by the private sector if provided the proper incentives.

The intent of the new law was to replace this situation with a "single, uniform national policy designed to cut down on bureaucracy and encourage private industry to utilize government financed inventions through the commitment of the risk capital necessary to develop such inventions to the point of commercial application."[3] Expanded technology commercialization was to be accomplished by employing the patent system to augment collaboration between universities (as well as other nonprofit institutions) and the business community to ensure that inventions are brought to market. The Bayh-Dole Act also provides for the increased participation of small firms in the national R&D enterprise under the assumption that these companies tend to be more innovative than larger companies.

The Patent System: A Brief Overview

The patent system was created to promote invention and innovation. Article I, Section 8, Clause 8 of the U.S. Constitution states: "The Congress Shall Have Power ... To promote the Progress of Science and useful Arts, by securing for limited Times to Authors and Inventors the exclusive Right to their respective Writings and Discoveries.... " Patents are widely believed to encourage innovation by simultaneously protecting the inventor and fostering competition. They provide the inventor with a right to exclude others, temporarily, from use of the invention without compensation. Patents give the owner an exclusive right for 20 years (from date of filing) to further develop the idea, commercialize a product or process, and potentially realize a return on the initial investment. Concurrently, the process of obtaining a patent places the concept in the public arena. As a disclosure system, the patent can, and often does, stimulate other firms or individuals to invent "around" existing patents to provide for parallel technical developments or meet similar market needs.[4]

Not everyone agrees that the patent system facilitates innovation. Critics argue that patents provide a monopoly which induces additional social costs. Others assert that the patent system is unnecessary due to market forces that already suffice to create an optimal level of innovation. The desire to obtain a lead time advantage over competitors, as well as the recognition that technologically backward firms lose out to their rivals, may well provide sufficient inducement to invent without the need for further incentives.[5] Some commentators believe that the patent system encourages industry concentration and presents a barrier to entry in some markets and that cross licensing between companies can result in exploitation of other markets.[6] Still other observers believe that the patent system too frequently attracts speculators who prefer to acquire and enforce patents rather than engage in socially productive activity.[7]

The importance of patents varies among industrial sectors. Patents are perceived as critical in the drug and chemical industries in part because of the ease of replicating the finished product. While it is expensive, complicated, and time consuming to duplicate an airplane, it is relatively simple to chemically analyze a pill and reproduce it.[8] Studies have found that in many other industries the protection offered by patents is diminished by the ability to invent around the patent and limited by the disclosure of vital information in the patent itself.[9] In the aircraft and semiconductor industries, patents have not been the most successful mechanism for capturing the benefits of investments.

Instead, lead time and the strength of the learning curve were determined to be more important.[10] Later studies bear this out; secrecy and lead time were deemed to have greater effect than patents in the semiconductor and related equipment industry, as well as the aerospace and machine tool industries, among others.[11]

Patents can provide an economic incentive for companies to pursue further development and commercialization. Studies indicate that research funding accounts for approximately one-quarter of the costs associated with bringing a new product to market. According to *The Economist*, "A dollar's worth of academic invention or discover requires upwards of $10,000 of private capital to bring [it] to market."[12] Patent ownership is seen as a way to encourage the additional, and often substantial investment necessary for new goods and services, particularly in the case of small business. In an academic setting, the possession of title to inventions is expected to provide motivation for the university to license the technology to the private sector for commercialization in anticipation of royalty payments.

University-Industry Cooperation

Changes to the patent laws embodied in the Bayh-Dole Act had as an objective the facilitation of collaborative ventures between and among academia, industry, and government. In 1980, universities performed 14% of the R&D undertaken in the United States (similar to today); much of this the fundamental research basic to technological advance.[13] The work is accomplished as part of the education process and provides training for scientists, engineers, and managers subsequently employed by the private sector.

Universities, however, generally do not have the means of production necessary to take the results of research and generate marketable products. Such activities are carried out by industry. Thus, the emphasis in the Bayh-Dole Act on the promotion of cooperative efforts between academia and the business community. By providing universities with intellectual property ownership with which to pursue and structure collaborative ventures, the legislation is intended to encourage the two sectors to work together to generate new goods, processes, and services for the marketplace. Such joint work allows for shared costs, shared risks, shared facilities, and shared expertise.

Prior to World War II, industry was the primary source of funding for basic research in universities. This financial support helped shape priorities and build relationships. However, after the war, the federal government supplanted the private sector as the major financial contributor and became the principal determinant of the type and direction of the research performed in academic institutions. This situation oftentimes resulted in a disconnect between the university and industrial communities. Because the private sector and not the government typically is involved in commercialization, the difficulties in moving an idea from the research stage to a marketable product or process appeared to have been compounded. Thus, efforts to encourage increased collaboration between and among the sectors through the Bayh-Dole Act were expected to augment the contribution of both parties to technological advancement.

Small Business

Special consideration concerning patent title was given to small businesses in part because of the role these companies were seen as playing in the generation of new jobs and in technological advancement. Early research supported by several federal agencies concluded that small, high technology companies are the source of significant innovation. An often cited 1982 study financed by the Small Business Administration determined that small firms were 2.4 times as innovative per employees as large companies.[14] Similar work performed at the time the legislation was being considered found that firms of less than 1,000 employees were responsible for more major innovations than large firms in the years 1953-1966 and for an equal number from 1967-1973.[15] More recent research points to the contribution of small businesses to economic growth "as measured by net new job creation.... "[16] According to the National Science Foundation, "U.S. small business is closely associated with the development of new technologies in many of the science-based industries likely to be important to future economic growth."[17]

Commentators argue that small firms act as entrepreneurs and change agents, undertaking innovative activities that stimulate the evolution of new and existing industries.[18] Because these companies often are involved in "leading-edge technical niches," the contribution of this sector to innovation "is most intense in new technologies."[19] Experts claim that

> small firm innovators are extremely effective at producing technically important innovations—and technically important innovations are more than twice as likely as large firm innovations to be extremely high impact.[20]

Therefore, small companies appear to be dominant in certain high technology industries including biotechnology, new materials, information technology and communications.[21]

However, certain caveats need to be stated particularly within the context of small business, innovation, and technology development. Over the years, experts have argued that the contribution of small firms to the economy is overstated. Marc Levinson, writing in *Dun's Business Month* during the 1980s, maintained that small companies tended to produce fewer goods than larger ones because they are less capital intensive and, on the whole, add less to the gross national product because they offer lower salaries and often do not provide health insurance or pension plans.[22] Professors Zoltan Acs and David Audretsch argued that the relationship between company size and innovation capacity varied by industry.[23] They note that "the evidence also suggests that there tends to be more innovative activity in industries consisting of larger and not smaller firms."[24] One commentator claims that the

> value of the innovations tended to increase with the size of the innovating firms. Consequently ... the interpretations that small firms are more innovative (or more efficient innovators) than large firms because they have introduced a larger number of innovations relative to their employment is unsound.[25]

Others maintain that there is no conclusive evidence that firm size affects the "success" of R&D.[26]

An important factor affecting the ability of small companies to effect technological advance appears to be the relationship between these firms and large corporations, a concept that is reflected in the provisions of the Bayh-Dole Act:

> the corporate contribution and that of the innovative entrepreneur are characteristically very different from one another and characteristically play complementary roles. Moreover, the contribution of the two together is superadditive, that is, the combined result is greater than the sum of their individual contributions.[27]

As small firms look to larger companies for additional resources, large firms look for partners as the new technologies developed by smaller companies look more viable.[28] Small businesses tend to be willing to take those technological risks that are not pursued by large firms and may be in a position to quickly exploit market opportunities. However, they may need to depend on large companies to meet large-scale manufacturing and/or market needs. In specific cases, experts note, "an innovative disadvantage of large firms is an innovative advantage for small firms, and vice versa, which can make collaboration between two firms of different size desirable for both parties."[29]

BAYH-DOLE ACT AND RELATED LAW

Provisions

In enacting P.L. 96-517, the Congress accepted the proposition that providing title to the contractor will encourage commercialization and that this should be used to support innovation in certain identified sectors. The law states:

> It is the policy and objective of the Congress to use the patent system to promote the utilization of inventions arising from federally-supported research or development; ... to promote collaboration between commercial concerns and nonprofit organizations, including universities; ... to promote the commercialization and public availability of inventions made in the United States by United States industry and labor; [and] to ensure that the Government obtains sufficient rights in federally-supported inventions to meet the needs of the Government and protect the public against nonuse or unreasonable use of inventions.... [30]

Each nonprofit organization (including universities) or small business is permitted to elect (within a reasonable time) to retain title to any "subject invention" made under federally funded R&D.[31] (According to a recent Supreme Court decision in *Stanford University v. Roche Molecular Systems Inc.*, "The Bayh-Dole Act does not automatically vest title to federally funded inventions in federal contractors or authorize contractors to unilaterally take title to such inventions." The act only clarifies "the order of priority of rights between the Federal Government and a federal contractor in a federally funded invention that already belongs to the contractor" and that certain conditions must be met before the invention belongs to the contractor.[32]) The institution must commit to commercialization within a predetermined, agreed upon, time frame. However, the government may keep title under "exceptional circumstances when it is determined by the agency that restriction or elimination of the right to retain title to any subject invention will better promote the policy and objectives of this chapter." Additionally, the government may withhold title if the contractor "is not located in the United States or does not have a place of business located in the United States or is subject to the control of a foreign government," in situations associated with national security, or when the work is related to the naval nuclear propulsion or weapons programs of the Department of Energy.[33]

Certain other rights are reserved for the government to protect the public's interests. The government retains "a nonexclusive, nontransferable, irrevocable, paid-up license to practice or have practiced for or on behalf of the United States any subject invention throughout the world.... " The government also retains "march-in rights" which enable the federal agency to require the contractor (whether the organization owns the title or has an exclusive license) to "grant a nonexclusive, partially exclusive, or exclusive license in any field of use to a responsible applicant or applicants ..." (with due compensation) or to grant a license itself under certain circumstances. The special situation necessary to trigger march-in rights involves a determination that the contractor has not made efforts to commercialize within an agreed upon time frame or that the "action is necessary to alleviate health or safety needs which are not reasonably satisfied by the contractor.... "[34]

The government is "authorized" to withhold public disclosure of information for a "reasonable time" until a patent application can be made. Licensing by any contractor retaining title under this act is restricted to companies which will manufacture substantially within the United States. Initially, universities were limited in the time they could grant exclusive licenses for patents derived from government sponsored R&D to large companies (5 of the *then* 17 years of the patent). This restriction, however, was voided by P.L. 98-620, the Trademark Clarification Act of 1984. According to S.Rept. 98-662, extending the time frame for licensing to large firms "is particularly important with technologies such as pharmaceuticals, where long development times and major investments are usually required prior to commercialization."[35]

Most experts continue to argue that patent exclusivity is important for both large and small firms. In a February 1983 memorandum concerning the vesting of title to inventions made under federal funding, then President Ronald Reagan ordered all agencies to treat, as allowable by law, all contractors regardless of size the same as prescribed in P.L. 96-517. This, however, does not have a legislative basis. P.L. 98-620, noted above, further amended Bayh-Dole by loosening the time limitations for both disclosure of an invention to the government agency and for the amount of time provided within which to elect to take title. Nonprofit institutions were subsequently permitted to assign title rights to another organization (e.g., one which markets technology) and government-owned, contractor-operated laboratories (primarily those of the Department of Energy) run by nonprofits were permitted to retain title to inventions made in the facility with the exception of those dedicated to naval nuclear propulsion or weapons development. In addition, the Federal Technology Transfer Act (P.L. 99-502) allows firms regardless of size to be awarded patents generated under a cooperative research and development agreement (CRADA) with a federal laboratory.[36]

Implementation and Results

The Bayh-Dole Act appears to have met its expressed goals of using "the patent system to promote the utilization of inventions arising from federally-supported research or development; ... and to promote collaboration between commercial concerns and nonprofit organizations, including universities.... "[37] In one of the earliest studies of the legislation, the General Accounting Office (now the Government Accountability Office, GAO) found agreement among university administrators and small business representatives that P.L. 96-517 had "a significant impact on their research and innovation efforts."[38] While noting it was not correct to generalize about academia from the 25 universities studied, GAO did find that by 1987 all university administrators questioned indicated that the Bayh-Dole Act had "been significant in stimulating business sponsorship of university research, which has grown 74 percent" from FY1980 to FY1985.[39] According to the National Science Foundation (NSF), industry support for academic research grew faster than any other funding source until FY2002. Industry financing expanded from 3.9% of university R&D in 1980 to 7.2% in 2000, although by FY2009 industry support had dropped to 5.8% of academic R&D. In 1980, federal financing comprised 67.5% of the total academic undertaking; by 2000 federal support declined to 58.2% of university funding, yet increased to 59.3% in FY2009.[40] It should be noted, however, that the federal government still remains the major source of academic research funding.

The majority of the university personnel involved in the GAO study indicated that the increase in industry support for research at universities was "directly" attributed to the patent changes in P.L. 96-517 and P.L. 98-620. Academic faculty interviews conducted by GAO found that "since businesses knew that universities could take title to federally funded inventions, they no longer were concerned that their research efforts could be 'contaminated' by federal funding with the possibility that a federal agency could assert title rights to resulting inventions."[41] All respondents agreed that the removal of licensing restrictions on nonprofit institutions (including universities) by P.L. 98-620 was of vital importance in promoting industry-university interaction.[42] This was reinforced by the finding that 9 out of 10 business executives questioned identified the Bayh-Dole Act as an "important factor" in their decisions to fund R&D in academia.[43]

Another GAO study published in May of 1998 reported that agency and university representatives believed the Bayh-Dole Act was meeting its goals as articulated by the Congress and the law had a positive impact on all involved. Academia was "receiving greater benefits from their inventions and were transferring technology better than the government did when it retained title to inventions."[44] In addition, the report states that the increased commercialization of federally funded research resulting from the implementation of the act positively affected both the federal government and the American people.[45]

Other experts agree. Yale President Richard Levin argued that the purpose of the Bayh-Dole Act is to transition the results of government funded research "into practice for the benefit of humanity" and that results indicate a "pretty emphatic positive answer that the Bayh-Dole Act has created public benefits" with minimal costs.[46] As stated in an article in *The Economist*, the BayhDole Act is "probably the most inspired piece of legislation to be enacted in America over the past half-century.... "[47]

One of the major factors in the reported success of the Bayh-Dole Act is the certainty it conveys concerning ownership of intellectual property. The Director of Stanford University's Office of Technology Licensing, Katherine Ku, noted that exclusivity is what motivates firms to invest financial and human resources in technology development.[48] It provides an incentive for universities to take the time and effort to pursue a patent and to license those patents in its portfolio. This has led to a significant increase in academic patenting. In 1980, 390 patents were awarded to universities;[49] by 2009, the number increased to 3,088.[50]

Academia has become a major source of innovation for local and regional economic development. In the latest survey (FY2011) performed by the Association of University Technology Managers (AUTM), universities identified a total of 591 new commercial products were marketed as a result of academic R&D. In addition, the survey indicated the creation of more than 671 new companies to commercialize university research with 6,051 new licenses/options granted primarily to small businesses. Since 1980, more than 8,778 new firms have been established to develop and market academic R&D, with "3,927 startups still operating as of the end of FY2011."[51]

Many of the start-up businesses initially created from university R&D were associated with just seven schools including the Massachusetts Institute of Technology (MIT), the University of California, California Tech, the University of Minnesota, the Johns Hopkins University, the University of Utah, and the University of Virginia.[52] While only a few universities earn large returns from licensing,[53] studies indicated that licensing by the University of California system generates $91 million in net licensing income annually with Columbia University receiving approximately $80 million and Florida State University $45 million.[54] Forbes also found that

> in many cases, a few big hits went a long way. New York University—which pulled in $175 million in research-related income on $210 million in research and development expenditures—tops the list with a 75% yield. Credit NYS's serious return in great part to smash-hit Remicade, a rheumatoid arthritis drug developed along with Centacor and Johnson & Johnson.[55]

A recent report found that "without accounting for product substitution effects, ... over the period 1996 to 2007, university licensing agreements based on product sales contributed at least $47 billion and as much as $187 billion to the U.S. GDP."[56]

However, several analysts argue that "Bayh-Dole was only one of a number of important factors behind the rise of university patenting and licensing activity."[57] In a study of the technology transfer and patenting activities of the University of California, Stanford University, and Columbia University, Professor David Mowery and his colleagues concluded that increased federal funding for basic biomedical research, expanded research in biotechnology, specific court rulings, and government policies augmenting what can be patented all contributed to the rise in academic intellectual property activities. According to their assessment, the Bayh-Dole Act had "little impact on the content of academic research." The pursuit of patenting and licensing at universities has expanded because of changes in biomedical and biotechnology R&D, not because of the act.[58] Later work by Professor Mowery follows this approach, arguing that "the emphasis on the Bayh-Dole Act as a catalyst to these interactions [increased university-industry cooperation and technology transfer] also seems somewhat misplaced, ignoring as it does the long history, extending to at least the earliest decades of the 20th century, of collaboration and knowledge flows between universities and industry in the United States."[59]

Some experts criticize this assessment and point out that the act had the most significant impact on universities that were not actively engaged in patenting prior to its passage.[60] Proponents of this position argue that as a result of the Bayh-Dole Act, in part, "university patenting increased particularly rapidly during the second half of the 1980s and early 1990s."[61] This growth in patenting has been concentrated in "middle-tier" schools, not just the top research universities.[62] Researchers who take this position suggest that the Mowery et al. study focused solely on universities that were previously involved in patenting and licensing and may not have fully considered patent problems that existed before the legislation was implemented. According to critics of the study, the analysts also failed to take into account changes in the venture capital industry that promoted the development of start-up companies to commercialize the results of university R&D.[63]

Other research questions the effect of increased university licensing on U.S. innovation. A study by Bhavan Sampat suggests that while the Bayh-Dole Act augmented patent and licensing by universities, these activities are just "one of many channels through which universities make economic contributions and in most industries less important contributions than those made by placing scientific and technological information in the public domain."[64] This author's work indicates that "there is little evidence that increased university patenting and licensing has facilitated increased technology transfer or any meaningful growth in the economic contributions of universities."[65]

However, commentators argue that the provisions of the Bayh-Dole Act provide incentives to take university inventions and develop them into products for the marketplace.[66] University technology generally is in the early stage and not yet ready for commercialization, requiring additional funding and the involvement of faculty to move the idea into a marketable product.[67] While most universities do not receive large amounts of funds as a result of licensing their technologies, it

> is clear from the evidence ... that faculty involvement in the further development of university technologies is an important element in getting those technologies to market. Mechanisms to ensure such efforts are an important element of commercialization regardless of whether those mechanisms included licensing by universities.[68]

In addition, Professor Scott Shane observes:

> Because universities exploit their inventions primarily through the licensing of technology, and licensing is not equally effective across all technologies ... the incentive to become more commercially focused led universities to concentrate their patenting in fields in which knowledge is transferred effectively through licensing.[69]

While the effects of the Bayh-Dole Act on the small business sector have not been as extensively studied, the results appear similar. All eight small business owners interviewed by GAO for its 1987 study indicated that the patent changes had a significant beneficial effect on research, development, and innovation in their firms.[70] Perhaps most illustrative of the influence of the Bayh-Dole Act on small business is the biotechnology industry. According to Dr. Bernadine Healy, the former Director of the National Institutes of Health (NIH), P.L. 96-517 was responsible for the development and growth of the biotechnology sector.[71] The biotechnology industry primarily is composed of small firms that are developing technologies and techniques derived from R&D funded by NIH. Many of these companies have been established by NIH alumni or university professors previously supported by NIH grants. In Senate testimony delivered on August 1, 2001, Dr. Marie Freire, then Director of the Office of Technology Transfer at NIH, stated that "[i]t is widely recognized that the Bayh-Dole Act and the Federal Technology Transfer Act continue to contribute to the global leadership of the U.S. biomedical enterprise.... " An industry that was in its infancy when the Bayh-Dole Act was passed, by the end of 2009 1,699 biotechnology firms generated annual sales of $48.2 billion.[72] The number of U.S. biotechnology patents granted has increased from 619 in 1985 to 4,853 in 2010.[73]

The value of the Bayh-Dole Act might be reflected in state efforts to promote industry-university cooperation based on the contributions of these activities to local economic growth. As Mark Myers, retired Senior Vice-President of Xerox, told a meeting of the National Academy of Sciences, "[t]he role of the research university is growing ever important as an economic force in our economy.... "[74] In a report by Battelle for the Biotechnology Industry Organization (BIO), analysts found that there are biotechnology-related initiatives in 44 states, including many that involve cooperative efforts between academia and the private sector. In 2008, 28 states and Puerto Rico had specific programs to facilitate industry-university collaboration in the biotechnology arena. Pre-seed and seed fund programs have been established in 25 states and Puerto Rico while 19 states have venture capital initiatives to invest in biotechnology R&D.[75] State laws also have been changed to allow universities to become equity partners in start up firms designed to commercialize academic R&D.[76] Later analysis indicates that by 2010, 42 states support initiatives to assist in the commercialization of new technologies in the bioscience arena.[77]

Current Issues and Concerns

While the Bayh-Dole Act provides a general framework to promote expanded utilization of the results of federally funded research and development, questions have been raised as to the adequacy of current arrangements. Most experts agree that closer cooperation among government, industry, and academia can augment funding sources (both in the private and public sectors), increase technology transfer, stimulate more innovation (beyond invention), lead to new products and processes, and expand markets. However, others point out that cooperation may provide an increased opportunity for conflict of interest, redirection of research, less openness in sharing of scientific discovery, and a greater emphasis on applied rather than basic research.

The successes of the Bayh-Dole Act and the visibility of the results of its implementation have generated certain concerns, many of which are associated with the role of the university in research, as well as biomedical and biotechnology R&D, particularly as related to the availability and cost of pharmaceuticals. Several of these issues are discussed below. However, it is important to place the Bayh-Dole Act in context. The law is one significant factor in expanded industry, university, small business collaboration, but not the only one. Therefore, it may be difficult to assess what concerns are the direct result of the Bayh-Dole Act and which arise from the overall research environment. The rising costs associated with the performance of research and development, the availability of venture capital, increased R&D outsourcing by large firms, and expanded federal funding for biomedical research all contribute to increased interaction among the parties. Additional legislative initiatives including the research and experimentation tax credit, the National Cooperative Research Act, the small business technology transfer program, the advanced technology program, and cooperative R&D agreements established by the StevensonWydler Technology Innovation Act all facilitate joint R&D activities leading to the commercialization of new technologies for the marketplace.[78]

Recoupment

Over the years, several legislators have suggested that the government "recoup" its investments from firms using federally supported research and development after profits are generated. This is particularly true in the area of pharmaceuticals.[79] Such arguments are similar to those that were identified and considered as part of the original legislative debate over patent policy and cooperative R&D. The concept of recoupment is based upon the argument that the government should be reimbursed for research and development expenses provided to a contractor if the resulting product is brought to the market and generates profits. Proponents of this approach also maintained that providing the contractor with a limited time monopoly on the results of federally funded R&D through assignment of patent rights should be balanced by compensation for the government's initial investment. In the debate over related legislation, then-Senator Robert Dole stated on the floor of the Senate on April 23, 1980, the provision for recoupment was intended to insure that "the Government's investment, paid for by the taxpayers of this country, is returned to the Federal coffer."[80] During the same debate, Senator Birch Bayh argued that a payback provision means that, "in the final analysis, the taxpayer will not be out the cost of the research and they also will have the benefit of the product."[81]

Such suggestions are based on several factors. In addition to funding research performed by individual companies, under certain circumstances, the government furnishes the private sector ownership of the intellectual property resulting from this public investment. Patent protection gives firms monopoly rights on these innovations for a specified amount of time. By providing patent protection to the results of federally funded research, a company receives an individual benefit based upon public investments. Thus, proponents of recoupment assert that the monopoly power of patents should be modified by this "public subsidization." They contend that the public has a right to a return on its investment. However, it is argued that "this right is not preserved under the patent system, which ascribes solely to the patent holder all proprietary rights and interests in the patented product or process."[82]

To date, Congress has weighed these issues and decided that, in the case of patent and technology policies, the benefits to the nation brought about by increased innovation are paramount. The passage of the Bayh-Dole Act represented a determination that, with respect to certain types of organizations, the economic incentive to realize a return on investment provided by a patent is necessary to stimulate companies to provide the often substantial financial commitment to turn federally funded R&D into marketable technologies and techniques. This is suggestive of the idea that the promise of a large return on investment "is precisely the tool sanctioned by the Constitution to promote the progress of science."[83] The decision was based on several determinations deriving from the rationale for federal support of basic research, the importance of technological progress to the nation, and the critical role of private sector commercialization in technological advancement.

Federal support for basic research is founded, in large part, on the understanding that the rate of return to society as a whole generated by investments in research is significantly larger than the benefits that can be captured by any one firm performing it.[84] It has been estimated that the returns to society generated by investments in basic research are approximately twice those to the company performing the work. Government support reflects a consensus that basic research is the foundation for many innovations, but that incentives for private sector financial commitments are dampened by the fact that spending for R&D runs a high risk of failure. Even results of fruitful R&D often are exploited by other domestic and foreign companies, thus resulting in underinvestment in research by the private sector. The returns from basic research are generally long term, sometimes not marketable, and not always evident.

It is now widely accepted that "from one-third to one-half of all [U.S.] growth has come from technical progress, and that it is the principal driving force for long-term economic growth and the increased standards of living of modern industrial societies."[85] Technological advancement can clearly contribute to the resolution of those national problems which are amenable to technological solutions. Such progress is achieved through innovation, the process by which industry provides new and improved products, processes, and services. An invention becomes an innovation when it has been integrated into the economy such that the knowledge created results in a new or improved good or service that can be sold in the marketplace or is applied to production to increase productivity and quality. It is only through commercialization, a function of the business sector, that a significant stimulus to economic growth occurs. Thus, there is congressional interest in accelerating development and commercialization activities in the private sector through the Bayh-Dole Act as well as other legislation.

Actual experience and cited studies suggest that companies which do not control the results of their investments—either through ownership of patent title, exclusive license, or pricing decisions—tend to be less likely to engage in related R&D. This likelihood is reflected in the provisions of the Bayh-Dole Act (as well as other laws). Providing universities, nonprofit institutions, and small businesses with title to patents arising from federally funded R&D offers an incentive for cooperative work and commercial application. Royalties derived from intellectual property rights provide the academic community an alternative way to support further research and the business sector a means to obtain a return on their financial contribution to the endeavor. While the idea of recoupment was considered by the Congress in hearings on the legislation, it was rejected as an unnecessary obstacle, one which would be perceived as an additional burden to working with the government. It was thought to be particularly difficult to administer.[86] Instead, Congress accepted as satisfactory the anticipated payback to the country through increased revenues from taxes on profits, new jobs created, improved productivity, and economic growth. For example, as discussed above, from 1980, when the Bayh-Dole Act was passed, through 2010, 8,107 new spin-off companies were created, and, in 2010 alone, 657 new products were introduced into the market by these firms. The emergence of the biotechnology industry and the development of new therapeutics to improve health care are other prominent indications of such benefits. To date, these benefits have been considered more important than the initial cost of the technology to the government or any potential unfair advantage.

Government Rights: Royalty Free Licenses and Reporting Requirements

As discussed above, the government retains certain rights under the Bayh-Dole Act to protect the public interest. The act states that the government is provided a "nonexclusive, nontransferable, irrevocable, paid-up license to practice or have practiced for or on behalf of the United States any subject invention throughout the world.... " This license, commonly known as a "royalty free license," has been the subject of some discussion including whether or not this permits government purchasers to obtain discounts on products developed from federally funded R&D, particularly pharmaceuticals. A July 2003 GAO report addressed this issue and concluded that the license entitles the government to practice or have practiced the invention on the government's behalf, but "does not give the federal government the far broader right to purchase, 'off the shelf' and royalty free (i.e. at a discounted price), products that happen to incorporate a federally funded invention when they are not produced under the government's license."[87] The study goes on to say that rights in one patent do not "automatically" permit rights in subsequent, related patents.[88] Because the government apparently holds few licenses on the biomedical products it purchases (generally through the Veteran's Administration and the Department of Defense),[89] federal officials indicated that procurement costs were best reduced by use of the Federal Supply Schedule and national contracts.[90] Government licenses are used primarily in the performance of research in the biomedical area.[91]

A related issue is that of tracking the government's interest in patents resulting from federally funded research and development. In an August 1999 study, GAO noted that federal contractors and grantees were not meeting the reporting requirements associated with the Bayh-Dole Act, making it difficult to identify and assess what licenses the government retained, among other things.[92] Two years later, in a follow-up report, GAO stated that four of the five agencies had taken steps to insure improved compliance with the law including several new monitoring systems, although more needed to be done.[93] Of particular interest is iEdison, created by the NIH, which electronically tracks federal inventions and is used by other agencies in addition to NIH.[94]

University Research

A question often posed is whether or not patent ownership rights provided by P.L. 96-517 have interfered with the traditional operating procedures of academia. A fear is that private sector funding of university R&D has led to conflicts of interest by scientists performing the research, particularly when academics have equity positions in the relevant companies. There are concerns that industry agendas will distort or supplant the basic research and educational responsibilities of academia. Complaints have also been expressed that the free exchange of ideas and scientific discovery are constrained as a result of both the university and the business community's interest in protecting their competitive positions.

The issue of conflict of interest is a complex one particularly when trying to determine what direct role the Bayh-Dole Act has in generating such concerns and what are the results of other factors that have lead to increased industrial funding of university research. As noted above, laws that provide tax incentives for private sector financing of university basic research and facilitate technology transfer and cooperative R&D among government, industry, and academia, as well as changes in the way companies obtain the basic research necessary for product development shape the environment within which academic research is pursued. Thus, as argued by Katherine Ku, it is necessary to evaluate criticisms of the Bayh-Dole Act and to understand that the success of the law has made many in government uncomfortable despite the clear guidelines for technology transfer it established.[95]

Senior Research Scholar Mildred Cho and her coauthors assert that the Bayh-Dole Act

> has created opportunities for conflict of interest for university faculty members because academic-industry partnerships can offer direct financial rewards to individual faculty members in the form of consulting fees, royalties, and equity in companies while simultaneously funding these faculty members' research.[96]

This, it is argued, has resulted in situations where the researcher's ties to private sector interests may not be evident and may adversely affect "the quality, outcome, and dissemination of research."[97] Other studies indicate that obligations to industry "pose a threat to scientific integrity."[98] Some commentators maintain that private sector funded research tends to generate conclusions favorable to industry; however, the factor that is primarily associated with the withholding or delay of information is the involvement of the scientist in bringing his research to market in a product, not the industrial financing itself.[99]

Data collected by Professor David Blumenthal and his colleagues also support the assessment that involvement in commercialization activities is related to delays in publication.[100] This study indicated that approximately 20% of life science researchers delayed publication of their studies more than six months at least once for reasons associated with patents and commercialization considerations. Almost 9% of faculty refused to share research or materials with other university scientists in the past three years. However, the authors conclude that "withholding of research results is not a widespread phenomenon among life-science researchers."[101] A survey of industry-university research centers by Professor Wesley Cohen and his colleagues found that over half of the centers permitted firms to request publication delays and 35% of the institutions allowed researchers to delete information prior to publication. At those centers with a mission to improve industrial products and processes, 63% allowed publication delays and 54% permitted the deletion of information.[102]

Delays in publication and the free flow of information from academia, according to Professor Richard Florida, "may well discourage or even impede the advancement of knowledge, which retards the efficient pursuit of scientific progress, in turn slowing innovation in industry."[103] Professor Florida also points to concerns over the increasing number of academic institutions taking equity positions in and/or incubating spin-off companies. These actions "simply tend to distract the university from its core missions of conducting research and generating talent." Florida concludes that publication delays and greater secrecy in the research process resulting from implementation of the Bayh-Dole Act have shifted the university away from the pursuit of its traditional goals.

Other experts, including Robert Barchi, Provost of the University of Pennsylvania, maintain that the Bayh-Dole Act has not generated a significant set of issues concerning conflicts of interest and publication delays primarily because of the importance of academic freedom to the faculty.[104] Publications are the basis for promotion and tenure and methods to respect reasonable intellectual property protection have been established. Similarly, as noted by Professor Pam Samuelson, conflicts of interest would jeopardize tenure thus regulations are in place to instruct faculty what is required of them.[105] Research conducted by Professors Pierre Azoulay, Waverly Ding, and Toby Stuart indicates that

> patenting is often accompanied by a flurry of publication activity ... academic scientists who patent are more productive than otherwise equivalent scientists that are not listed as inventors on patents, but that publication quality appears relatively similar in the two groups.[106]

In response to these issues, many universities have hired professional technology managers to work with faculty and to address patents. Universities with extensive research capabilities and resources were the first to create offices of technology transfer; after passage of the Bayh-Dole Act these offices were established with much greater frequency.[107] These university technology transfer offices have established guidelines to cover industry-university relationships, with education and publication remaining academic priorities.[108] The financial rewards derived from patenting often are only a small portion of the total amount of R&D funding for academic institutions and what substantial money does flow into individual institutions tends to be the result of one "blockbuster" patent. University technology managers report that the major reason for patent licensing is commercialization, not profit, particularly since the cost of a patent, which can run approximately $10,000, is so high.[109] While the Bayh-Dole Act focused universities on "commercially relevant technologies and closer ties between research and technological development,"[110] the costs of patenting are such that "most university licensing offices barely break even."[111]

University limitations on outside research, expeditious publication obligations mandated for certain federally funded R&D, and conflict of interest provisions also help to preserve a balance between federal policies like the Bayh-Dole Act that promote industry-university cooperation and concerns over excessive control of the research environment by the business community. For example, NIH requires grant recipients to publish the results of their government funded R&D. This is augmented by tax code regulations necessitating prompt dissemination of actual research results in order for a university or research institution to retain its tax exempt status. NIH also has policies and guidelines promoting the availability of patents arising from federal funding for use by other scientists for research purposes without acquisition of a license.[112]

Critics argue that the Bayh-Dole Act is distorting the traditional role of the university to the detriment of future technological development. Professor Florida maintained that because universities are seen as "engines" of growth, they focus on applied rather than fundamental research. This has lead to unrealistic national and local policies and practices that encourage the commercialization of academic research while ignoring the real value of universities as the "nation's primary source of knowledge creation and talent."[113] Mildred Cho also asserted that university research is "skewed" toward marketable products and not basic research.[114] Studies by researchers Dianne Rahm and Robert P. Morgan et. al. indicated the greater the faculty interaction with industry the more the applied research.[115] According to an article in *Fortune* magazine, the Bayh-Dole Act has had "unintended consequences" in that "universities have evolved from public trusts into something closer to venture capital firms. What used to be a scientific community of free and open debate now often seems like a litigation scrum of data-hoarding and suspicion."[116]

Other experts disagree. A study of 3,400 faculty at six major research institutions by Professors Jerry Thursby and Marie Thursby found that "the basic/applied split in research did not change over the period 1983-1999 even though licensing had increased by a factor greater than 10."[117] Data collected by the National Science Foundation appear to support this assessment. According to NSF, in 1980, basic research comprised 66.6% of academic R&D endeavors while applied research and development were 33.4% of the total. In 2009, the percent of academic R&D expenditures devoted to basic research increased to 74.6% while applied research and development declined to 25.4% of the total.[118]

Commentators claim that the Bayh-Dole Act encourages the type of research that is attractive to faculty. James Severson, President of the Cornell Research Foundation, testified before the House Committee on the Judiciary that

> Today, the protection and commercialization of academic research is one way for universities to attract, retain, and reward talented faculty who wish to see the results of their research programs benefit society. A commitment to the protection of research results is important for universities to develop closer ties to companies, and to attract additional funds to support research programs.[119]

As noted by Terry Young, Assistant Vice Chancellor for Technology Transfer at Texas A&M University, the act requires funds generated by licensing to be used for future education and research necessary to "deliver 'real world' products to the public."[120] Assessing the legislation, the Biotechnology Industries Association, contends that "without the Bayh-Dole Act, few licensing agreements would be executed between private companies and federally supported research institutions, and the enormous investment our government makes in medical research would be wasted."[121]

Biotechnology and Pharmaceuticals

Many of the current concerns about the Bayh-Dole Act primarily arise out of its application to the biotechnology and pharmaceutical industries.

Congressional interest in providing lower cost drugs, particularly to seniors, has focused attention on the role the act has had on the development of new pharmaceuticals for the marketplace. Certain critics maintain that the price of many therapeutics derived from federally funded R&D are excessive considering the government's financial contribution.[122] Others argue that the Bayh-Dole Act does not significantly affect pharmaceutical prices and point to a July 2001 study by NIH that found only 4 of the 47 FDA approved drugs generating $500 million a year were developed in part with NIH funded technologies.[123] Although the government generally does not directly support pharmaceutical research aimed at product development,[124] legislative attempts have been made to require cost controls or recoupment on drugs generated, in part, with federal funds. This is in sharp contrast to congressional and executive branch efforts, particularly in the defense arena, to make it easier for firms to acquire and utilize intellectual property associated with federally financed R&D.[125]

Overall support for biological and medical sciences has grown significantly since the passage of the Bayh-Dole Act. As measured in constant 2000 dollars, total (federal and non-federal) spending for academic R&D in these areas has increased from $4.6 billion in 1980 to $22.1 billion in 2008.[126] Funding for university R&D in the life science, particularly biological and medical sciences, comprises by far the largest portion of academic research support. In 2008, 52.1% of total R&D expenditures at academic institutions went to finance the medical and biological sciences. When the Bayh-Dole Act was passed in 1980, 40.5% of the research spending at universities was in these areas.[127] Expanded support for university R&D in this arena appears to be important in relation to findings by the late Professor Edwin Mansfield showing that academic research was particularly significant in the development of new products and processes in the pharmaceutical and medical device industries.[128]

Interest and activity in the biomedical and biotechnology sectors has sparked some concern over the effects of the Bayh-Dole Act on research in these areas. According to information provided by the Boston Consulting Group, in the years between 1990 and 1999, new gene patents granted increased from about 400 to 2,800 while the number granted to universities expanded from 55% to 73% during that time period.[129] Another study "estimated that in the U.S. over 3,000 new DNA-related patents have issued every year since 1998, and more than 40,000 such patents have been granted."[130] Similarly, the number of U.S. biotechnology patents granted continues to grow.[131] The focus on intellectual property has led critics to charge that the Bayh-Dole Act encourages the patenting of fundamental research which, in turn, prevents further biomedical innovation. Law professors Rebecca Eisenberg and Arti Rai argue that due to the legislation, "proprietary claims have increasingly moved upstream from the end products themselves to the ground-breaking discoveries that made them possible in the first place."[132] While patents are designed to spur innovation, Rai and Eisenberg maintain that certain patents hinder the process. From their perspective, by permitting universities to patent discoveries made under federal funding, the Bayh-Dole Act "draws no distinction between inventions that lead directly to commercial products and fundamental advances that enable further scientific studies."[133] These basic innovations are generally known as "research tools."

Eisenberg and Professor Richard Nelson argue that ownership of research tools may "impose significant transaction costs" that result in delayed innovation and possible future litigation.[134] It also can stand in the way of research by others:

> Broad claims on early discoveries that are fundamental to emerging fields of knowledge are particularly worrisome in light of the great value, demonstrated time and again in history of science and technology, of having many independent minds at work trying to advance a field. Public science has flourished by permitting scientists to challenge and build upon the work of rivals.[135]

Similar concerns were expressed by Harold Varmus, President of Memorial Sloan-Kettering and former Director of NIH. In July 2000 prepared testimony, he spoke to being "troubled by widespread tendencies to seek protection of intellectual property increasingly early in the process that ultimately leads to products of obvious commercial value, because such practices can have detrimental effects on science and its delivery of health benefits."[136] While the Bayh-Dole Act and scientific advances have helped generate a dynamic biotechnology industry, there have been changes that "are not always consistent with the best interests of science."[137]

However, as Varmus and others acknowledge, the remedies to this situation are not necessarily associated with the Bayh-Dole Act. Yale's Richard Levin noted that while some research should be kept in the public domain, including research tools, the fact that it is privatized is not the result of the Bayh-Dole Act, but rather the result of patent law made by the courts and the Congress. Therefore, he believes that changes to the act are not the appropriate means to address the issues.[138] Other experts agree that "many of the issues that are identified today as negative consequences of Bayh-Dole can be traced to the institutional polices [of universities] structured to optimize institutional benefits and income, rather than to the Act itself."[139]

Current law, as reaffirmed by court decisions, permits the patenting of research tools. However, there have been efforts to encourage the widespread availability of these tools. Marie Freire testified that the value to society is greatest if the research tools are easily available for use in research. She asserted that there is a need to balance commercial interests with public interests.[140] To achieve this balance, NIH has developed guidelines for universities and companies receiving federal funding that make clear research tools are to be made available to other scientists under reasonable terms.[141] In addition, the U.S. Patent and Trademark Office recently made changes in the guidelines used to determine the patentability of biotechnology discoveries.

Studies by Professors John Walsh, Ashish Arora, Wesley Cohen, and Charlene Cho found that although there are now more patents associated with biomedical research, and on more fundamental work, there is little evidence that work has been curtailed due to intellectual property issues associated with research tools.[142] Scientists are able to continue their research by "licensing, inventing around patents, going offshore, the development and use of public databases and research tools, court challenges, and simply using the technology without a license (i.e., infringement)." According to the authors, private sector owners of patents permitted such infringement in academia (with the exception of those associated with diagnostic tests in clinical trials) "partly because it can increase the value of the patented technology."

CONCLUSION

The discussion surrounding changes to the patent laws in the 1980s, and the debate over technology transfer since the late 1970s, acknowledged many of the issues currently being explored. As a result of expressed concerns, certain safeguards were built into the activities authorized by the Bayh-Dole Act. As discussed previously, march-in rights, the government's retention of an irrevocable license to patents generated under federally funded R&D, publication requirements, and commercialization schedules, among other things, all are incorporated into the process to protect the public interest. While there is a potential for creating an "unfair" advantage for one company over another, this is balanced against the need for new technologies and techniques and their contribution to the well-being of the nation.

Despite arguments that title should remain in the public sector where it is accessible to all interested parties, the earlier lack of exclusivity appeared to interfere with the further development and commercialization of federally funded R&D. During the 1980s, Congress determined that the dispensation of patent rights to universities, small businesses, and nonprofit institutions and cooperative efforts took precedence, projecting the greater good generated by new products and processes that improve the country's health and welfare. Lawmakers anticipated the economic benefits through increased revenues from profits, wages, and salaries. The government receives a significant payback through taxes on profits and society benefits from new jobs created and expanded productivity. The importance of patent ownership has been reinforced by the positive effects studies have demonstrated P.L. 96-517 is reported to have had on the emergence of new technologies and new techniques generated by American companies.

There remain areas of concern, as discussed above, that Congress may decide to pursue. Some argue, particularly with respect to pharmaceuticals and biotechnology, that under the Bayh-Dole Act companies are receiving too many benefits at the expense of the public. Others, particularly in the defense arena, assert that the existing rights retained by the government under the act are too restrictive and are an impediment to meeting federal needs. But the impact of the legislation is still seen as significant. As summed up by Howard Bremer, who was patent counsel to the Wisconsin Alumni Research Foundation from 1960 through 1988:

> One important factor, which is often overlooked, is that the success was achieved without cost to the taxpayer. In other words, no separate appropriation of government funds was needed to establish or manage the effort. In fact, it has been estimated that the economic benefits flowing from the universities' licensing activities adds about $41 billion to the United States economy.
>
> Significant as that dollar amount is, it should not be overlooked that university inventions, arising, as most of them do, from basic research, have led to many products which have or exhibit the capability of saving lives or of improving the lives, safety and health of the citizens of the United States and around the world. In that context their contribution to society is immeasurable.[143]

End Notes

[1] For additional discussion, see CRS Report RL33528, Industrial Competitiveness and Technological Advancement: Debate Over Government Policy, by Wendy H. Schacht.

[2] National Science Board, Science and Engineering Indicators—2006, (Washington, National Science Foundation, 2006), A4-5.

[3] House Committee on the Judiciary, Report to Accompany H.R. 6933, 96th Cong., 2d Sess., H.Rept. 96-1307, Part 1, 3.

[4] For more information, see CRS Report RL32324, Federal R&D, Drug Discovery, and Pricing: Insights from the NIH-University-Industry Relationship, by Wendy H. Schacht.

[5] Frederic M. Sherer, Industrial Market Structure and Economic Performance (Rand McNally & Co, 1970), 384-87.

[6] John R. Thomas, "Collusion and Collective Action in the Patent System: A Proposal for Patent Bounties," University of Illinois Law Review, 2001, 305.

[7] Ibid.

[8] Federic M. Scherer, "The Economics of Human Gene Patents," Academic Medicine, December 2002, 1350.

[9] Wesley M. Cohen, Richard R. Nelson, and John P. Walsh, Protecting Their Intellectual Assets: Appropriability Conditions and Why U.S. Manufacturing Firms Patent (or Not), National Bureau of Economic Research, February 2000, available at http://www.nber.org /papers/w7552.

[10] Richard C. Levin, Alvin K. Klevorick, Richard R. Nelson, and Sidney G. Winter, "Appropriating the Returns for Industrial Research and Development," Brookings Papers on Economic Activity, 1987, printed in The Economics of Technical Change, Edwin Mansfield and Elizabeth Mansfield, eds., (Vermont, Edward Elgar Publishing Co., 1993), 253.

[11] Protecting Their Intellectual Assets: Appropriability Conditions and Why U.S. Manufacturing Firms Patent (or Not), Table 1.

[12] "Innovation's Golden Goose," The Economist (US), December 14, 2002, available at http://www.economist.com/ node/1476653?story_id=1476653.

[13] National Science Board, Science and Engineering Indicators—2002 (Washington, National Science Foundation, 2002), A4-9.

[14] National Science Board, Science and Engineering Indicators—1993 (Washington, National Science Foundation, 1993), 185.

[15] National Science Board, Science Indicators—1976 (Washington, National Science Foundation, 1976), 116.

[16] BJK Associates, The Influence of R&D Expenditures on New Firm Formation and Economic Growth, Small Business Administration, Office of Advocacy, October 2002, 2, available at http://www.njit.edu/v2/News/Releases/ finalreport_10-02-02.pdf.

[17] National Science Foundation, "Indicators of U.S. Small Business's Role in R&D," InfoBrief, NSF 10-304, March 2010, 1, available at http://www.nsf.gov/statistics /infbrief /nsf10304/nsf10304.pdf.

[18] Martin A. Carree and A. Roy Thurik, "Entrepreneurship, Economic Growth and Policy," in Handbook of Entrepreneurship Research, Zoltan J. Acs and David B. Audretsch, eds. (New York, Springer Press, 2003), 439.

[19] CHI Research, Inc, Small Serial Innovators: The Small Firm Contribution to Technological Change, Small Business Administration, Office of Advocacy, February 2003, 3, available at http://www.sba.gov/advo/research/rs225tot.pdf.

[20] Ibid., 12.

[21] CHI Research, Inc., Small Firms and Technology: Acquisitions, Inventor Movement, and Technology Transfer, Small Business Administration, Office of Advocacy, January 2004, 4, available at http://www.sba.gov/advo/research/ rs233tot.pdf.

[22] Marc Levinson, "Small Business: Myth and Reality," Dun's Business Month, September 1985, 32-33.

[23] Zoltan J. Acs and David B. Audretsch, Innovation and Small Firms (Cambridge, The MIT Press, 1990), 50-51.

[24] Ibid., 147.

[25] B.S. Tether, "Small and Large Firms: Sources of Unequal Innovation?," Research Policy, November 1998, 742.

[26] Charles Brown, James Hamilton, and James Medoff, Employers Large and Small, (Cambridge, Harvard University Press, 1990), 10.

[27] William J. Baumol, Education for Innovation: Entrepreneurial Breakthroughs vs. Corporate Incremental Improvements, NBER, June 2004, 2-3, available at http://www.nber.org/papers /10578.

[28] David R. King, Jeffrey G. Covin, W. Harvey Hegarty, "Complementary Resources and the Exploitation of Technological Innovations," Journal of Management, August 1, 2003, 595.

[29] Ibid., 592.

[30] 35 U.S.C. §200.

[31] 35 U.S.C. §202.

[32] Board of Trustees of the Leland Stanford Junior University v. Roche Molecular Systems, Inc., et al., __U.S.__(June 6, 2011).

[33] 35 U.S.C. §202.

[34] 35 U.S.C. §203.

[35] Senate Committee on the Judiciary, Report to Accompany S. 2171, 98th Cong., 2d Sess. S.Rept. 98-662, 1984, 3.

[36] For additional discussion see Industrial Competitiveness and Technological Advancement: Debate Over Government Policy.

[37] 35 U.S.C. §200.

[38] U.S. General Accounting Office, Patent Policy: Recent Changes in Federal Law Considered Beneficial, RCED-87- 44, April 1987, 3.

[39] Ibid., 3.

[40] National Science Foundation, "Changes in Federal and Non-Federal Support for Academic R&D Over the Past Three Decades," InfoBrief, June 2002, available at http://www.nsf.gov, National Science Foundation, National Patterns of R&D Resources: 2003, Special Report, available at http://www.nsf.gov/statistics/nsf05308/pdfstart.htm, and National Science Foundation, "Universities Report $55 Billion in Science and Engineering R&D Spending for FY2009; Redesigned Survey to Launch in 2010." InfoBrief, NSF 10-329, 1, available at http://www.nsf.gov/statistics/infbrief/ nsf10329/nsf10329.pdf.

[41] Patent Policy: Recent Changes in Federal Law Considered Beneficial, 20-21.

[42] Ibid., 16.

[43] Ibid., 23.

[44] U.S. General Accounting Office, Technology Transfer: Administration of the Bayh-Dole Act by Research Universities, RCED-98-126, May 1998, 2.

[45] Ibid., 15.

[46] National Academy of Sciences, Board on Science, Technology, and Economic Policy, Workshop on Academic IP: Effects of University Patenting and Licensing on

Commercialization and Research, April 17, 2001 [transcript], 261- 262, available at http://www.nas.edu.

[47] Innovation's Golden Goose.

[48] Workshop on Academic IP: Effects of University Patenting and Licensing on Commercialization and Research, 9.

[49] Science and Engineering Indicators—1993, 430.

[50] National Science Board, Science and Engineering Indicators, 2012 (Washington, National Science Foundation, 2010), Appendix table 5-48, available at http://www.nsf.gov /statistics/seind12/append/c5/at05-48.pdf.

[51] Association of University Technology Managers, AUTM U.S. Licensing Activity Survey Highlights: FY2011, available at http://www.autm.net/AM/Template.cfm? Section=FY_ 2011_Licensing_Activity_Survey&Template=/.CM/ContentDisplay.cfm&ContentID=8731.

[52] Goldie Blumenstyk, "Income From University Licenses on Patents Exceeded $1-Billion," The Chronicle of Higher Education, March 22, 2002.

[53] Harun Bulut and GianCarlo Moschini, U.S. Universities' Net Returns from Patenting and Licensing: A Quantile Regression Analysis, Iowa State University Working Paper-06-WP 432, September 2006, 2, available at http://www.card.iastate.edu.

[54] Gregory K. Sobolski, John H. Barton, Ezekiel J. Emanuel, "Technology Licensing, Lessons From the US Experience," Journal of the American Medical Association, December 28, 2005, 3138.

[55] Maureen Farrell, "Universities That Turn Research into Revenue," Forbes.com, September 12, 2008, available at http://www.forbes.com/2008/09/12/google-general-electric-ent-tech-cx_mf_0912universitypatent.html.

[56] David Roessner, Jennifer Bond, Sumiye Okubo, and Mark Planting, The Economic Impact of Licensed Commercialized Inventions Originating in University Research, 1996-2007, Final Report to the Biotechnology Industry Organization, September 3, 2009, 32, available at http://www.bio.org/ip/techtransfer/ BIO_final_report_9_3_09_rev_2.pdf.

[57] David C. Mowery, Richard R. Nelson, Bhaven N. Sampat, and Arvids A. Ziedonis, "The Growth of Patenting and Licensing by U.S. Universities: An Assessment of the Effects of the Bayh-Dole Act of 1980," Research Policy 30, 2001, 99.

[58] Ibid., 100.

[59] David C. Mowery, The Bayh-Dole Act and High-Technology Entrepreneurship in U.S. Universities: Chicken, Egg, or Something Else?, paper prepared for the Eller Center conference on "Entrepreneurship Education and Technology Transfer," University of Arizona, January 21-22, 2005, available at http://entrepreneurship.eller.arizona.edu/docs/ conferences/2005/colloquium/D_Mowery.pdf.

[60] Workshop on Academic IP: Effects of University Patenting and Licensing on Commercialization and Research, 17.

[61] Science and Engineering Indicators—1993, 152.

[62] Ibid., 152 and Workshop on Academic IP: Effects of University Patenting and Licensing on Commercialization and Research, 57-58.

[63] Ashley J. Stevens, "Is Bayh-Dole Under Siege Again?" Technology Access Report, July 2001. See also Lori TurkBicakci and Steven Brint, "University-Industry Collaboration: Patterns of Growth for Low- and Middle-Level Performers," Higher Education, January 2005, 61-89.

[64] Bhavan N. Sampat, "Patenting and US Academic Research in the 20th Century: The World Before and After BayhDole," Research Policy, July 2006, 773.

[65] Ibid.

[66] Marie Thursby, Jerry Thursby, and Emmanuel Dechenaux, Shirking, Shelving, and Sharing Risk: The Role of University License Contracts, April 9, 2004, National Bureau of Economic Research, available at http://www.nber.org/ ~confer/2004/entf04/thursby.pdf.

[67] Jerry G. Thursby and Marie C. Thursby, University Licensing Under Bayh-Dole: What are the Issues and Evidence?, May 2003, available at http://opensource.mit.edu/papers /Thursby.pdf.

[68] Ibid.

[69] Scott Shane, "Encouraging University Entrepreneurship? The Effect of the Bayh-Dole Act on University Patenting in the United States," Journal of Business Venturing, 2004, 128.

[70] Patent Policy: Recent Changes in Federal Law Considered Beneficial, 4.

[71] House Committee on the Judiciary, Biotechnology Development and Patent Law, 102d Cong., 1st Sess., November 20, 1991, 48.

[72] Ernst & Young, Beyond Borders, Global Biotechnology Report 2010, 59, available at http://www.ey.com/Publication/vwLUAssets/Beyond_borders_2010/$FILE/Beyond_ borders_2010.pdf.

[73] Science and Engineering Indicators, 2010, Appendix table 6-58, available at http://www.nsf.gov/statistics/seind10/appendix.htm and Science and Engineering Indicators, 2012, Appendix table 6-60, available at http://www.nsf.gov/ statistics/seind12/append/c6/at06-60.pdf.

[74] Workshop on Academic IP: Effects of University Patenting and Licensing on Commercialization and Research, 255.

[75] Battelle Technology Partnership Practice, Technology, Talent and Capital: State Bioscience Initiatives 2008, Biotechnology Industries Organization, June 2008, available at http://bio.org/local/battelle2008/ State_Bioscience_Initiatives_2008.pdf.

[76] Battelle Technology Partnership Practice and SSTI, Laboratories of Innovation: State Bioscience Initiatives 2004, June 2004, 27-29, available at http://www.washington lifescience.com/econ_dev_reports/ OregonBioscienceInitiatives2004.pdf.

[77] Peter M. Pellerito and George Goodno, "Successful State Initiatives That Encourage Bioscience Industry Growth," Biotechnology Industry Organization, February 9, 2012, available at http://www.bio.org/node/5771.

[78] For additional information, see CRS Report RL33526, Cooperative R&D: Federal Efforts to Promote Industrial Competitiveness, by Wendy H. Schacht, and CRS Report RL33528, Industrial Competitiveness and Technological Advancement: Debate Over Government Policy, by Wendy H. Schacht.

[79] For a more detailed discussion of this issue in the pharmaceutical arena, see CRS Report RL32324, Federal R&D, Drug Discovery, and Pricing: Insights from the NIH-University-Industry Relationship, by Wendy H. Schacht.

[80] U.S. Congress, Congressional Record, April 23, 1980, S739.

[81] Ibid., S743.

[82] Steven R. Salbu, "Aids and Drug Pricing: In Search of a Policy," Washington University Law Quarterly, Fall 1993, 5-20.

[83] Evan Ackiron, "Patents for Critical Pharmaceuticals: The AZT Case," American Journal of Law and Medicine, 1991, 18.

[84] Edwin Mansfield, "Social Returns From R&D: Findings, Methods, and Limitations," Research/Technology Management, November-December 1991, 24.

[85] Gregory Tassey, The Economics of R&D Policy (Connecticut, Quorum Books, 1997), 54. See also: Edwin Mansfield, "Intellectual Property Rights, Technological Change, and Economic Growth," in: Intellectual Property Rights and Capital Formation in the Next Decade, eds.

Charls E. Walker and Mark A. Bloomfield (New York, University Press of America, 1988), 5.

[86] For example see U.S. House of Representatives, Committee on Science and Technology, Government Patent Policy, Hearings, September 23, 27, 28, 29, and October 1, 1976, 94th Cong. 2nd sess., 1976; United States Senate, Select Committee on Small Business, Government Patent Policies, Hearings, December 19, 20, and 21, 1977, 95th Cong. 1st sess., 1978; and U.S. Senate, Committee on Commerce, Science, and Transportation, Patent Policy, Hearings, July 23 and 27, and October 25, 1979, 96th Cong. 1st sess., 1979.

[87] General Accounting Office, Technology Transfer: Agencies' Rights to Federally Sponsored Biomedical Inventions, GAO-03-536, July 2003, 7.

[88] Ibid., 8.

[89] Ibid.

[90] Ibid., 12.

[91] Ibid., 10.

[92] General Accounting Office, Technology Transfer: Reporting Requirements for Federally Sponsored Inventions Need Revision, August 1999, GAO/RCED-99-242, 2.

[93] General Accounting Office, Intellectual Property: Federal Agency Efforts in Transferring and Reporting New Technology, October 2002, GAO-03-47, 29.

[94] Ibid., 33.

[95] Workshop on Academic IP: Effects of University Patenting and Licensing on Commercialization and Research, 98, 100-101.

[96] Mildred K. Cho, Ryo Shohara, Anna Schissel, and Drummond Rennie, "Policies on Faculty Conflicts of Interest at U.S. Universities," Journal of the American Medical Association, November 1, 2000.

[97] Ibid.

[98] Justin E. Bekelman, Yan Li, and Cary P. Gross, "Scope and Impact of Financial Conflicts of Interest in Biomedical Research: A Systematic Review," Journal of the American Medical Association, January 22/January 29, 2003.

[99] Ibid.

[100] David Blumenthal, Eric G. Campbell, Melissa S. Anderson, Nancyanne Causino, and Karen Seashore Louis, "Withholding Research Results in Academic Life," Journal of the American Medical Association, April 16, 1997, 1224.

[101] Ibid.

[102] Wesley M. Cohen, Richard Florida, Lucien Randazzese, and John Walsh, "Industry and the Academy: Uneasy Partners in the Cause of Technological Advance," in: Challenges to Research Universities, eds. Linda R. Cohen, Wesley Cohen, Roger Noll, William Rogerson, and Albert Teich (Washington, The Brookings Press, 1998), 188-189.

[103] Richard Florida, "The Role of the University: Leveraging Talent, Not Technology," Issues in Science and Technology, Summer 1999.

[104] Workshop on Academic IP: Effects of University Patenting and Licensing on Commercialization and Research,19- 20.

[105] Ibid., 193.

[106] Pierre Axoulay, Waverly Ding, and Toby Stuart, The Impact of Academic Patenting on the Rate, Quality, and Direction of (Public) Research, National Bureau of Economic Research, January 2006, available at http://www.nber.org/papers/w11917.pdf.

[107] Everett Rogers, Jing Yin, and Joern Hoffmann, "Assessing the Effectiveness of Technology Transfer Offices at U.S. Research Universities," Journal of the Association of University Technology Managers, v. XII, 2000, available at http://www.autm.net.

[108] Technology Transfer: Administration of the Bayh-Dole Act by Research Universities.

[109] Ann M. Thayer, "University Technology Moves to Market via Patenting, Licensing,: Chemical and Engineering News, August 24, 1992, 17-18. See also: Jerry G. Thursby and Marie C. Thursby, "Intellectual Property: University Licensing and the Bayh-Dole Act," Science, August 22, 2003, 1052.

[110] Science and Engineering Indicators—2002, 5-54.

[111] Lita Nelson, "Increase of Intellectual Property Licensing at Universities Stems from Changes in Funding and Legislation," MIT Tech Talk, August 26, 1998, available at http://web.mit.edu.

[112] Policies available at http://www.nih.gov.

[113] The Role of the University: Leveraging Talent, Not Technology.

[114] Eric Niller, "Biotech & Health: Report Fails to Address the Downside of Academic-Industry Collaborations," Wall Street Journal (Europe), August 6, 2001, 17.

[115] Industry and the Academy: Uneasy Partners in the Cause of Technological Advance, 186.

[116] Clifton Leaf, "The Law of Unintended Consequences," Fortune, September 19, 2005, 252.

[117] University Licensing Under Bayh-Dole: What are the Issues and Evidence? See also Jerry G. Thursby and Marie C. Thursby, "Has the Bayh-Dole Act Compromised Basic Research?" Research Policy, Vol. 40, Issue 8, October 2011, 1077-1084.

[118] Universities Report $55 Billion in Science and Engineering R&D Spending for FY2009; Redesigned Survey to Launch in 2010, 1.

[119] House Committee on the Judiciary, Subcommittee on Courts and Intellectual Property, Hearings on Gene Patents and Other Genomic Inventions, July 13, 2000, available at http://www.house.gov/judiciary/seve0713.htm.

[120] U.S. Department of Commerce, Technology Administration, Innovation in America: University R&D, June 11, 2002, available at http://www.ta.doc.gov/reports.

[121] Biotechnology Industry Organization, Testimony on Bayh-Dole and Technology Transfer Before the President's Council on Science and Technology, Office of Science and Technology Policy, April 11, 2002, available at http://www.bio.org.

[122] See CRS Report RL32324, Federal R&D, Drug Discovery, and Pricing: Insights from the NIH-University-Industry Relationship, by Wendy H. Schacht.

[123] Department of Health and Human Services, National Institutes of Health , NIH Response to the Conference Report Request for a Plan to Ensure Taxpayers' Interests are Protected, July 2001, available at http://www.nih.gov/news/ 070101wyden.htm.

[124] See CRS Report RL30913, Pharmaceutical Research and Development: A Description and Analysis of the Process, by Richard E. Rowberg.

[125] See House Committee on Government Reform, Subcommittee on Technology and Procurement Policy, Toward Greater Public-Private Collaboration in Research and Development: How the Treatment of Intellectual Property Rights is Minimizing Innovation in the Federal Government, hearings, July 17, 2001, available at http://www.house.gov /reform.

[126] Science and Engineering Indicators, 2010, Appendix table 5-6, available at http://www.nsf.gov/statistics/seind10/ append/c5/at05-06.pdf.

[127] Ibid.

[128] Edwin Mansfield, "Academic Research and Industrial Innovation: An Update of Empirical Findings," Research Policy, 1998, 773-776.

[129] Hamilton Moses, III and Joseph B. Martin, "Academic Relationships with Industry," Journal of the American Medical Association, February 21, 2001, 933.

[130] Timothy Caulfield, Robert M. Cook-Deegan, F. Scott Kieff, and John P. Walsh, Evidence and Anecdotes: An Analysis of Human Gene Patenting Controversies, Nature Biotechnology, September 2006, available at http://www.ncbi.nlm.nih.gov/pmc/articles/PMC2701726/.

[131] Science and Engineering Indicators, 2010, Appendix table 6-58.

[132] Arti K. Rai and Rebecca S. Eisenberg, "Bayh-Dole Reform and the Progress of Biomedicine," American Scientist, January- February 2003, 52.

[133] Ibid.

[134] Rebecca S. Eisenberg and Richard R. Nelson, "Public vs. Proprietary Science: A Fruitful Tension?," Daedalus, Spring 2002.

[135] Ibid.

[136] Hearings on Gene Patents and Other Genomic Inventions.

[137] Ibid.

[138] Workshop on Academic IP: Effects of University Patenting and Licensing on Commercialization and Research, 262.

[139] Sara Boettiger and Alan B. Bennet, "Bayh-Dole: If We Knew Then What We Know Now," Nature Biotechnology, 2006, available at http://www.nature.com/nbt/journal/v24 /n3/full/nbt0306-320.html.

[140] Senate Committee on Appropriations, Subcommittee on Labor, Health and Human Services, Education and Related Agencies, Hearings, August 1, 2001.

[141] Available on the NIH website at http://www.nih.gov.

[142] John P. Walsh, Ashish Arora, Wesley M. Cohen, "Working Through the Patent Problem," Science, February 14, 2003, 1021 and John P. Walsh, Charlene Cho, and Wesley Cohen, "View for the Bench: Patents and Material Transfers," Science, September 23, 2005, 2002-2003.

[143] Howard Bremer, "The First Two Decades of the Bayh-Dole Act as Public Policy," National Association of State Universities and Land-Grant Colleges, November 11, 2001, available at http://www.nasulgc.org.

INDEX

D

E

F

G

H

I

T

U

V

W

Y